Jamal Berakdar

Electronic Correlation Mapping

From Finite to Extended Systems

Related Titles

Berakdar, J.

Concepts of Highly Excited Electronic Systems

315 pages with 13 figures
2003
Hardcover
ISBN 3-527-40335-3

Berakdar, J.

SET:

I. Concepts of Highly Excited Electronic Systems
II. Electronic Correlation Mapping - From Finite to Extended Systems

approx. 565 pages in 2 volumes
2006
Hardcover
ISBN 3-527-40412-0

Berakdar, J., Kirschner, J. (Eds.)

Correlation Spectroscopy of Surfaces, Thin Films, and Nanostructures

255 pages with 132 figures
2004
Hardcover
ISBN 3-527-40477-5

Jamal Berakdar

Electronic Correlation Mapping

From Finite to Extended Systems

WILEY-VCH Verlag GmbH & Co. KGaA

The Author

Jamal Berakdar
MPI für Mikrostrukturphysik Halle, Germany
e-mail: jber@mpi-halle.de

Cover picture

by Jamal Berakdar

Library of Congress Card No.:
applied for

British Library Cataloguing-in-Publication Data
A catalogue record for this book is available from the British Library.

Bibliographic information published by Die Deutsche Bibliothek
Die Deutsche Bibliothek lists this publication in the Deutsche Nationalbibliografie; detailed bibliographic data is available in the Internet at <http://dnb.ddb.de>.

Typesetting Uwe Krieg, Berlin
Printing Strauss GmbH, Mörlenbach
Binding J. Schäffer Buchbinderei GmbH, Grünstadt

Printed in the Federal Republic of Germany
Printed on acid-free paper

ISBN-13: 978-3-527-40350-9
ISBN-10: 3-527-40350-7

To my family

Contents

1 Qualitative and General Features of Electron–Electron Scattering **1**

1.1 Mapping Momentum-distribution Functions 1

1.2 Role of Momentum Transfer during Electron–Electron Scattering 5

1.3 Approximate Formula for the Electron–Electron Ionization Cross Section . . 8

1.3.1 Example: An Atomic Target 11

1.3.2 Electron–Electron Cross Section for Scattering from Condensed Matter 13

1.3.3 Electron Scattering Cross Section from Ordered Materials 14

1.3.4 Initial- vs. Final-state Interactions 15

1.4 Averaged Electron–Electron Scattering Probabilities 15

1.4.1 Integrated Cross Section for Strongly Localized States 16

1.4.2 Low-energy Regime 19

1.5 Electron–Electron Scattering in an Extended System 19

2 Spin-effects on the Correlated Two-electron Continuum **23**

2.1 Generalities on the Spin-resolved Two-electron Emission 24

2.2 Formal Symmetry Analysis 27

2.3 Parametrization of the Spin-resolved Cross Sections 29

2.4 Exchange-induced Spin Asymmetry 30

2.5 Physical Interpretation of the Exchange-induced Spin Asymmetry 32

2.6 Spin Asymmetry in Correlated Two-electron Emission from Surfaces 33

2.7 General Properties of the Spin Asymmetry 35

2.7.1 Spin Asymmetry in Pair Emission from Bulk Matter 36

2.7.2 Spin-polarized Homogenous Electron Gas 37

Electronic Correlation Mapping: From Finite to Extended Systems. Jamal Berakdar

ISBN: 3-527-40350-7

2.7.3 Behavior of the Exchange-induced Spin Asymmetry in Scattering from Atomic Systems 37
2.7.4 Threshold Behavior of the Spin Asymmetry 39

3 Mechanisms of Correlated Electron Emission 43
3.1 Exterior Complex Scaling 44
3.2 The Convergent Close Coupling Method 45
3.3 Analytical Models 46
3.3.1 Dynamical Screening 47
3.3.2 Influence of the Density of Final States 48
3.4 Analysis of the Measured Angular Distributions 51
3.4.1 The Intermediate Energy Regime 51
3.5 Characteristics of the Correlated Pair Emission at Low Energies 53
3.5.1 Influence of the Exchange Interaction on the Angular Pair Correlation 55
3.6 Threshold Behavior of the Energy and the Angular Pair Correlation 57
3.6.1 Generalities of Threshold Pair Emission 57
3.6.2 Threshold Pair Emission from a Coulomb Potential 59
3.6.3 Regularities of the Measured Pair Correlation at Low Energies 60
3.6.4 Role of Final-state Interactions in Low-energy Correlated Pair Emission 62
3.6.5 Interpretation of Near-threshold Experiments 63
3.7 Remarks on the Mechanisms of Electron-pair Emission from Atomic Systems 70

4 Electron–electron Interaction in Extended Systems 73
4.1 Exchange and Correlation Hole 74
4.2 Pair-correlation Function 76
4.2.1 Effect of Exchange on the Two-particle Probability Density 78
4.3 Momentum-space Pair Density and Two-particle Spectroscopy 79
4.3.1 The S Matrix Elements 79
4.3.2 Transition Probabilities and Cross Sections 81
4.3.3 Two-particle Emission and the Pair-correlation Function 82

5 The Electron–Electron Interaction in Large Molecules and Clusters **85**

5.1 Retardation and Nonlocality of the Electron–Electron Interaction in Extended Systems . . . 86

5.2 Electron Emission from Fullerenes and Clusters . . . 90

5.2.1 The Spherical Jellium Model . . . 91

5.2.2 Angular Pair Correlation . . . 92

5.2.3 Total Cross Sections . . . 95

5.2.4 Finite-size Effects . . . 96

5.2.5 Influence of Exchange . . . 97

6 Pair Emission from Solids at Surfaces **101**

6.1 Qualitative Analysis . . . 102

6.1.1 Model Crystal Potential . . . 103

6.1.2 Scattering from the Surface Potential . . . 105

6.1.3 Qualitative Features of Interacting Two-particle Emission from Surfaces . . . 106

6.1.4 Explicit Results for Two-particle Scattering from Metal Surfaces . . . 107

6.2 Mechanisms of Correlated Electron Emission . . . 111

6.2.1 Angular Pair Correlation . . . 111

6.2.2 Energy Pair Correlation . . . 114

6.2.3 Influence of Exchange on the Energy Pair Correlation . . . 117

6.2.4 Pair Diffraction . . . 120

6.3 Role of the Dynamical Collective Nature of the Two-particle Interaction . . . 123

6.4 Quantitative Description of Pair Emission from Surfaces . . . 126

6.4.1 Treating Strong Two-particle Correlations . . . 127

6.4.2 Relativistic Layer KKR Method . . . 130

6.4.3 Two-particle Energy Correlation in the Pair Emission from Tungsten . 132

6.4.4 Angular Pair Correlation: Role of the Electron–Electron Interaction . 133

7 Pair Emission from Alloys **135**

7.1 Correlated Two-particle Scattering from Binary Substitutional Alloys 136

7.1.1 Pair Emission from Alloys in Transmission Mode . . . 136

7.1.2 Pair Emission in Reflection Mode 137
7.1.3 Scattering Potential from Binary Alloys 138
7.1.4 Electronic States and Disorder Averaged Spectral Functions 139
7.2 Incorporation of Damping of the Electronic States 140
7.3 Configurationally Averaged Cross Section 142
7.3.1 Analytical Model for Configurationally Averaged Cross Section . . . 144
7.4 Numerical Results and Illustrations 146

Color Figures **149**

Appendices

A Electronic States in a Periodic Potential **155**

B Screening Within Linear Response Theory **159**
B.1 Kubo Formalism 159
B.2 Density–density Correlation Functions 160

C Lindhard Function **163**
C.1 Thomas–Fermi Approximation 164
C.2 Friedel Oscillations 165
C.3 Plasmon Excitations 165

D Dynamic Structure Factor and the Pair-distribution Function **167**
D.1 Excitation Processes and the Dynamical Structure Factor 169
D.2 Properties of the Pair-distribution Function 171

References **173**

Index **187**

Preface

In many-particle spectroscopy a sample is disturbed by an approaching test particle with well-defined properties. The response of the system depends decisively on the amount of energy and momentum transferred during the interaction with the external particle. If allowed by energy- and momentum-conservation laws, the sample's constituents may be excited into continuum states where their quantum numbers can be resolved simultaneously with those of the testing particle. Scanning the reaction probability as a function of these quantum numbers one can investigate various aspects that may be categorized as follows: (i) One can study the single-particle spectrum of the sample and how it is modified by the presence of other degrees of freedom. (ii) One can assess the frequency and the momentum-dependent potential experienced by the test particle when it couples to the target. (iii) Momentum-space few-particle correlation functions can also be accessed. Due to its inherent many-body character, the many-particle coincidence or correlation spectroscopy is uniquely suited for the study of the latter two categories. Originally, the coincidence technique was developed and utilized to address problems in nuclear physics, and soon after the method was applied to atomic, molecular and condensed-matter systems. It is however only recently that the coincidence spectroscopic techniques have been advanced and refined to a level where their unique features can be fully exploited, in particular as far as the mapping of properties akin to correlated many-body systems is concerned.

This experimental advance is paralleled with equally important progress in the numerical and conceptual understanding of the physics of correlated finite and extended systems. The aim of the present treatise is not to review or even summarize all of the experimental and theoretical achievements in this field but to distill some simple but general mechanisms of how interacting electronic systems respond to an external perturbation induced by an incident particle. In a further step we reflect on how these mechanisms and the system's response can

Electronic Correlation Mapping: From Finite to Extended Systems. Jamal Berakdar

ISBN: 3-527-40350-7

serve as a tool to map out the properties of interparticle correlations by means of many-particle spectroscopies. Only a small fraction of the available body of experimental and theoretical findings is employed for this purpose and I apologize to all the colleagues whose works and results could not be addressed and/or fully analyzed here.

The book starts with a qualitative analysis of the outcome of the two-particle correlation spectroscopy of localized and delocalized electronic systems, as they occur in atoms and solids. The second chapter addresses how spin-dependent interactions can be imaged by means of correlation spectroscopy and points out similarities and differences between finite spin-polarized systems (such as polarized atoms) and extended systems (ferromagnets).

A further chapter discusses possible pathways for the production of interacting two-particle continuum states and provides illustration and analysis using some of the available experimental data on atomic systems. In addition, we explore to what extent these mechanisms remain viable when the system size grows.

To connect to known concepts in condensed-matter physics we present briefly in a further chapter some established ways of quantifying electronic correlations and point out the relationship to correlation spectroscopy.

Furthermore, we address in a separate chapter how the frequency and the momentum-transfer dependent response of an electronic system can be assessed by means of two-particle spectroscopy and illustrate the ideas by some applications to fullerenes and metal clusters.

The last two chapters are devoted to the investigation of the potential of two-particle spectroscopy in studying ordered surfaces and disordered samples. In particular, we explore the main possible processes for the generation of interacting two-particle states at surfaces and draw some conclusions as to what novel information is extractable from this spectroscopy.

Throughout the book the material is analyzed using rather qualitative arguments and the results of more sophisticated numerical theories serve the purpose of endorsing the suggested physical scenarios. The foundations of some of these theories have been presented in a first volume of this book.

Many of the ideas, conclusions and results presented in this book materialized over years of collaborations and in discussions with a number of friends and colleagues to whom I wish to express my thanks here. I am particularly indebted to L. Avaldi, I. Bray, J. S. Briggs, P. Bruno, R. Dörner, A. Dorn, R. Dreizler, A. Ernst, J. Feagin, R. Feder, N. Fominykh, H. Gollisch,

J. Henk, A. Kheifets, O. Kidun, J. Kirschner, H. Klar, K.A. Kouzakov, A. Lahmam-Bennani, J. Lower, D. Madison, S. Mazevet, R. Moshammer, A.R.P. Rau, J.-M.Rost, S. N. Samarin, H. Schmidt-Böcking, G. Stefani, A. T. Stelbovics, J. Ullrich, and E. Weigold.

Finally, it is my pleasure to thank Frau C. Wanka from Wiley-VCH for the precious help and support during the preparation of the manuscript.

Jamal Berakdar

Halle, August 2005

1 Qualitative and General Features of Electron–Electron Scattering

This chapter provides an outline of the qualitative features of the electron emission from localized (atomic) electronic states and delocalized states induced by electron impact. The analysis assumes a short scattering time. This means that the time τ_0 spent by an electron incoming with a velocity v_0 in the region $\mathcal{R}$ where the scattering potential U is active is small on the typical time scale τ within which U may have a significant influence. $\mathcal{R}$ may be quantified by some range a so that $\tau_0 \approx a/v_0$ and if U_0 is a typical strength of the potential energy then $\tau \approx \hbar/U_0$. Some special features pertinent to the infinite-range of the Coulomb potential will be discussed later.

1.1 Mapping Momentum-distribution Functions

We consider the following scenario: a continuum electron with a momentum $\hbar\boldsymbol{k}_0$ and energy E_0 scatters from an another electron that resides in a localized (atomic) state $\varphi(\boldsymbol{r}_2)$ formed in the field of a residual ion with a charge Z and a mass M (cf. the schematic drawing in Fig. 1.1). The binding energy is denoted by ϵ_i. Upon collision, the two electrons emerge with momenta $\boldsymbol{p}_1 = \hbar\boldsymbol{k}_1$ and $\boldsymbol{p}_2 = \hbar\boldsymbol{k}_2$. The recoil-ion momentum we denote by $\hbar\boldsymbol{k}_{\text{rec}}$.

Neglecting corrections of the order of $1/M$ we can evaluate[1] the fully-differential cross section (normalized to the incoming electron flux density) as [1–3]

$$\mathrm{d}\sigma(\boldsymbol{k}_1,\boldsymbol{k}_2,\boldsymbol{k}_{\text{rec}}) = (2\pi)^4\frac{1}{v_0}\,|T|^2\,\delta(E_{\text{f}}-E_{\text{i}})\delta(\boldsymbol{P}_{\text{f}}-\boldsymbol{P}_{\text{i}})\,\mathrm{d}^3\boldsymbol{k}_1\mathrm{d}^3\boldsymbol{k}_2\mathrm{d}^3\boldsymbol{k}_{\text{rec}} \qquad (1.1)$$

$$\mathrm{d}\sigma(\boldsymbol{k}_1,\boldsymbol{k}_2) = (2\pi)^4\frac{1}{v_0}\,|T|^2\,\delta(E_{\text{f}}-E_{\text{i}})\,\mathrm{d}^3\boldsymbol{k}_1\mathrm{d}^3\boldsymbol{k}_2 \qquad (1.2)$$

[1] Unless stated otherwise atomic units (a.u.) are used throughout the book, i.e., $\hbar$ is the unit of action, the Bohr radius a_0 is the unit of length. The units of charge and mass are those of the electron. The unit of energy is one Hartree which is the electrostatic potential energy of two electrons separated by one a_0.

Electronic Correlation Mapping: From Finite to Extended Systems. Jamal Berakdar

ISBN: 3-527-40350-7

where the matrix elements of the transition operator are given by

$$T = \langle \Psi_{\boldsymbol{k}_1\boldsymbol{k}_2}(\boldsymbol{r}_1, \boldsymbol{r}_2)|U|\varphi(\boldsymbol{r}_2)\psi_{\boldsymbol{k}_0}(\boldsymbol{r}_1)\rangle \tag{1.3}$$

Here $\Psi_{\boldsymbol{k}_1\boldsymbol{k}_2}(\boldsymbol{r}_1, \boldsymbol{r}_2)$ is the wave function of the two electrons in the final state, $\psi_{\boldsymbol{k}_0}(\boldsymbol{r}_1)$ is a plane wave describing the continuum electron in the initial state and U is the interaction potential; $\boldsymbol{r}_1$ and $\boldsymbol{r}_2$ are the position vectors of the two electrons measured with respect to the residual ion. The total energies in the initial and in the final channel E_i and E_f are

$$E_\mathrm{i} = E_0 + \epsilon_i, \quad E_\mathrm{f} = E_1 + E_2 + E_\mathrm{rec} \tag{1.4}$$

The conservation law of the total linear momentum in the initial ($\boldsymbol{P}_\mathrm{i}$) and final ($\boldsymbol{P}_\mathrm{f}$) channel dictates that

$$\boldsymbol{P}_\mathrm{i} = \boldsymbol{k}_0 = \boldsymbol{k}_1 + \boldsymbol{k}_2 + \boldsymbol{k}_\mathrm{rec} = \boldsymbol{P}_\mathrm{f} \tag{1.5}$$

The kinetic energy of the recoiling ion E_rec is generally small due to the large mass and usually has no influence on the structure of the cross sections. Assuming free motion of the final state electrons, i.e.

$$\Psi_{\boldsymbol{k}_1\boldsymbol{k}_2}(\boldsymbol{r}_1, \boldsymbol{r}_2) \approx (2\pi)^{-3} \exp(\mathrm{i}\boldsymbol{k}_1 \cdot \boldsymbol{r}_1 + \mathrm{i}\boldsymbol{k}_2 \cdot \boldsymbol{r}_2) \tag{1.6}$$

and that $U(\boldsymbol{r}_1, \boldsymbol{r}_2, \epsilon_i) = U(\boldsymbol{r}_1 - \boldsymbol{r}_2, \epsilon_i)$ we conclude from Eq. (1.2) that

$$\mathrm{d}\sigma(\boldsymbol{k}_1, \boldsymbol{k}_2) = 2\pi \frac{1}{v_0} \left|\tilde{U}(\boldsymbol{q})\right|^2 \left|\tilde{\varphi}(\boldsymbol{q} - \boldsymbol{k}_2)\right|^2 \delta(E_\mathrm{f} - E_\mathrm{i})\, \mathrm{d}^3\boldsymbol{k}_1 \mathrm{d}^3\boldsymbol{k}_2 \tag{1.7}$$

where $\boldsymbol{q} = \boldsymbol{k}_0 - \boldsymbol{k}_1$ is the momentum transfer. $\tilde{U}(\boldsymbol{q})$ and $\tilde{\varphi}(\boldsymbol{q} - \boldsymbol{k}_2) = \tilde{\varphi}(\boldsymbol{k}_\mathrm{rec})$ are, respectively, the Fourier transforms of the scattering potential and of the initially bound state.

Equation (1.7) makes clear what information can be extracted from studying the cross section at large momenta of the continuum electrons (in which case Eq. (1.6) is justified).

When the scattering potential U is known one may fix the value of $\boldsymbol{q}$ (by choosing fixed $\boldsymbol{k}_0$ and $\boldsymbol{k}_1$) and then vary $\boldsymbol{k}_2$ to map out the electronic density distribution in momentum space[2] $|\tilde{\varphi}(\boldsymbol{q} - \boldsymbol{k}_2)|^2$. This has been done successfully for a variety of target materials [4–6].

On the other hand formally, in a many-electron system the effective (two-particle) interaction U, which is generally not known in closed analytical form, depends in a characteristic way on the properties of the target, as detailed in Appendix B and discussed in depth in later

[2]This quantity is also called in the literature the X-ray form factor.

chapters. Hence, measuring the cross section (1.7) offers an insight into the form factor of the effective potential U.

The generic behavior of the cross section (1.2) is deducible from the simplified Eq. (1.7): Generally speaking, for $q \approx 0$ (distant collisions) $E_0 + \epsilon_i \approx E_1$ and hence k_2 and $|\boldsymbol{q} - \boldsymbol{k}_2|$ are small compared to k_1, as illustrated in Fig. 1.1. In this case the low-momentum Fourier components of φ are decisive for the cross section (1.2). For example, let us consider the electron scattering from an electron bound in the hydrogen atom. The electron–electron scattering potential is $U = 1/|\boldsymbol{r}_1 - \boldsymbol{r}_2|$ and $\varphi(\boldsymbol{r}_2) = \varphi_{nlm}(\boldsymbol{r}_2)$; hereby the hydrogenic orbitals are characterized by the principle n, orbital l and magnetic m quantum numbers. Hence, $|\tilde{U}(\boldsymbol{q})|^2 = \frac{2}{\pi\, q^4}$ and $\tilde{\varphi}(\boldsymbol{k}_{\text{rec}}) = N_{nlm} Y_{lm}(\hat{\boldsymbol{k}}_{\text{rec}}) F_{n,l}(k_{\text{rec}})$. Here $Y_{lm}(\hat{\boldsymbol{k}}_{\text{rec}})$ are spherical harmonics and N_{nlm} are normalization constants. The meaning of the m independent functions $F_{n,l}(k_{\text{rec}})$ is that $|k_{\text{rec}}\ F_{n,l}(k_{\text{rec}})|^2$ represents the *momentum distribution* for the state characterized by the quantum numbers n, l. Hence, the probability that the absolute magnitude of the expectation values of the momentum operator lies between k_{rec} and $k_{\text{rec}} + \Delta k_{\text{rec}}$ (regardless of the direction $\hat{\boldsymbol{k}}_{\text{rec}}$) is given by $|k_{\text{rec}} F_{n,l}(k_{\text{rec}})|^2 \Delta k_{\text{rec}}$ ($\int_0^\infty |k_{\text{rec}} F_{n,l}(k_{\text{rec}})|^2 \mathrm{d}k_{\text{rec}} = 1$). For a hydrogenic system $F_{n,l}(k_{\text{rec}})$ are known analytically, e.g., if we express k_{rec} in units of $Z\hbar/a_0$ (a_0 is the Bohr radius) we find

$$F_{1,0}(k_{\text{rec}}) = \frac{2^{5/2}}{\sqrt{\pi}} \frac{1}{(k_{\text{rec}}^2 + 1)^2}, \quad F_{2,0}(k_{\text{rec}}) = \frac{32}{\sqrt{\pi}} \frac{4k_{\text{rec}}^2 - 1}{(4k_{\text{rec}}^2 + 1)^3}, \tag{1.8}$$

$$F_{n,0}(0) = (-1)^{n-1} \sqrt{\frac{32}{\pi}} n^{5/2}, \tag{1.9}$$

$$\begin{aligned} F_{2,1}(k_{\text{rec}}) &= \frac{128}{\sqrt{3\pi}} \frac{k_{\text{rec}}}{(4k_{\text{rec}}^2 + 1)^3}, \quad F_{3,1}(k_{\text{rec}}) = \frac{864}{\sqrt{3\pi}} \frac{k_{\text{rec}}(9k_{\text{rec}}^2 - 1)}{(9k_{\text{rec}}^2 + 1)^4}, \\ F_{3,2}(k_{\text{rec}}) &= \frac{5184}{\sqrt{15\pi}} \frac{k_{\text{rec}}^2}{(9k_{\text{rec}}^2 + 1)^4} \end{aligned} \tag{1.10}$$

For a hydrogenic target and within the approximation (1.6) we can now deduce from Eqs. (1.7) and (1.8)–(1.10) the structure of the cross section as a function of $\hat{\boldsymbol{k}}_2$ and k_2 (note $\boldsymbol{k}_2 = \boldsymbol{q} - \boldsymbol{k}_{\text{rec}}$): Considering the dependence on k_2 (or E_2) we remark that the form factor of the scattering potential behaves as $|\tilde{U}(\boldsymbol{q})|^2 \sim q^{-4}$. Hence, for a fixed $\boldsymbol{q}$, the cross-section angular dependence is encapsulated solely in $Y_{lm}(\hat{\boldsymbol{k}}_{\text{rec}}) F_{n,l}(\boldsymbol{k}_{\text{rec}})$. E.g., for $n = 1, l = 0$ the cross section has a maximum at $\hat{\boldsymbol{k}}_2 \parallel \hat{\boldsymbol{q}}$ (cf. 1.8). This peak is conventionally referred to as the binary peak, even though its structure is determined by the initial bound state momentum distribution. For example, for $\boldsymbol{k}_2 = \boldsymbol{q}$ and $n = 2, l = 1$ we deduce from Eq. (1.10) that the

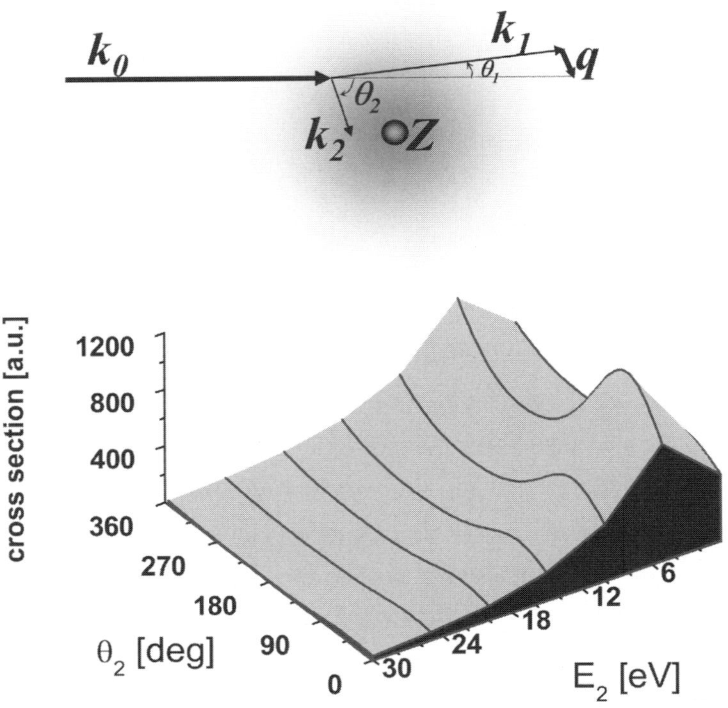

Figure 1.1: Top panel: a schematic view of the process under consideration: An electron with wave vector $\boldsymbol{k}_0$ scatters from an another electron which is bound to a residual ion with a charge Z. In the exit channel both electrons are in the continuum states characterized by the wave vectors $\boldsymbol{k}_1$ and $\boldsymbol{k}_2$. Lower panel: The fully differential cross section (1.7) for the process shown in the upper panel as a function of the energy E_2 and the polar angle θ_2 (with respect to $\hat{\boldsymbol{k}}_0$): The bound electron resides initially in the ground state of atomic hydrogen. The electron incident with an energy of 1 keV is scattered in the final state through an angle of $\theta_1 = -1°$. The final-state wave function is modelled by (1.6).

binary peak appears as a double peak because $F_{2,1}$ has a node at $\boldsymbol{k}_2 = \boldsymbol{q} \Rightarrow \boldsymbol{k}_{\mathrm{rec}} = 0$. Generally, the nodes in the momentum-distribution functions are directly reflected in zero points in the cross section. For s states $\tilde{\varphi}$ is finite at $\boldsymbol{k}_2 = \boldsymbol{q}$ ($k_{\mathrm{rec}} = 0$) and therefore the binary peak occurs when $\boldsymbol{k}_2$ coincides with $\boldsymbol{q}$ and behaves as a function of n, as dictated by Eq. (1.9). Having made this statement one should recall however that in Eq. (1.8) and (1.10) k_{rec} is measured in units of $Z\hbar/a_0$, meaning that if $Z \gg |\boldsymbol{q} - \boldsymbol{k}_2|$ then the binary peak is smeared out. This is clearly seen for the state $n = 1, l = 0$ (cf. Eq. (1.8)) where $F_{1,0} \rightarrow constant$ for

$|\boldsymbol{q} - \boldsymbol{k}_2|/Z \rightarrow 0$. The physical meaning of this is that as Z increases the particle location-uncertainty $\langle \Delta r \rangle$ decreases (as a_0/Z). Correspondingly, the momentum uncertainty increases as $\sim \hbar/\langle \Delta r \rangle \sim \hbar Z/a_0$ which sets the scale for the momentum spread of the momentum distribution function. Hence to observe a fully developed binary peak the quantity $|\boldsymbol{q} - \boldsymbol{k}_2|$ has to vary in a range of the order of or larger than $\hbar Z/a_0$. These arguments are applicable generally to the (stable) ground state of a potential acting within the range a: a rough estimate of the binary peak width is given by $\hbar/a$ which is a measure of the extent of the momentum-distribution function. Conversely a sharp binary peak with a width $< \Delta\kappa >$ is associated with a state in the configuration space extended roughly over the region $\hbar/ < \Delta\kappa >$.

The position and the structure of the binary peak is also associated with the structure of the spectrum and can also be roughly understood from a general point of view: For simplicity let us consider a one dimensional smooth confining (real) potential $V(x)$. The Schrödinger equation dictates that the bound-state wave function at the total energy ϵ_i should behave as $\partial_x^2 \varphi(x) = -\kappa^2 \varphi(x)$ where $\kappa^2(x) = \frac{2m}{\hbar^2}\left[\epsilon_i - V(x)\right]$. Hence, in the classically allowed regime (i.e. where the kinetic energy $\epsilon_i - V(x)$ is positive) the curvature of φ always has the opposite sign to φ and vanishes (i.e., changes sign) when $\varphi = 0$. Therefore, in the classically allowed regime $\varphi(x)$ exhibits oscillating behavior, regardless of the shape of $V(x)$. The (angular) frequencies ω_j of these oscillations show up as peaks in the momentum-distribution function $\tilde{\varphi}$ and hence the "binary" peak fragments in structures appearing at the wave vector $|\boldsymbol{q} - \boldsymbol{k}_2| = (\frac{2m\omega_j}{\hbar})^{1/2}$. We note in this context that to capture high-frequency oscillation large k_{rec} are required, which means large momentum transfer q. To ensure the validity of (1.6) one should also choose k_2 to be large. In addition, it should be noted that the energies of the initial state follow from (1.4) once the energies of the continuum electrons are determined.

1.2 Role of Momentum Transfer during Electron–Electron Scattering

To improve on the approximation (1.6) for the final state one can assume the initially bound electron to be ejected into a scattering state $\psi_{\boldsymbol{k}_2}(\boldsymbol{r}_2)$ of the target, whilst the impinging electron is still being described by plane waves in the initial and in the final channel. In the atomic physics literature this approximation is called the first Born approximation (FBA). Within the

FBA $|\psi_{\boldsymbol{k}_2}\rangle$ and $|\varphi\rangle$ are orthogonal and Eq. (1.2) reads in this case

$$\mathrm{d}\sigma(\boldsymbol{k}_1,\boldsymbol{k}_2) = 2\pi\frac{1}{v_0}\left|\tilde{U}(\boldsymbol{q})\right|^2\left|\tilde{\chi}(\boldsymbol{q},\boldsymbol{k}_2)\right|^2\delta(E_\mathrm{f}-E_\mathrm{i})\,\mathrm{d}^3\boldsymbol{k}_1\mathrm{d}^3\boldsymbol{k}_2 \tag{1.11}$$

where $\tilde{\chi}(\boldsymbol{q},\boldsymbol{k}_2)$ is

$$\tilde{\chi}(\boldsymbol{q},\boldsymbol{k}_2) = (2\pi)^{3/2}\int \mathrm{d}^3\boldsymbol{p}\,\tilde{\psi}^*_{\boldsymbol{k}_2}(\boldsymbol{p})\,\tilde{\varphi}(\boldsymbol{q}-\boldsymbol{p}) \tag{1.12}$$

Here $\tilde{\psi}_{\boldsymbol{k}_2}(\boldsymbol{p})$ is the Fourier transform of $\psi_{\boldsymbol{k}_2}(\boldsymbol{r}_2)$.

For the scattering kinematics explained in Fig. 1.1 we show, in Fig. 1.2a, the cross section calculated in the FBA. As is clear from this figure the main feature in the cross section introduced by FBA is the appearance of a peak at $\boldsymbol{k}_2 = -\boldsymbol{q}$. This peak is usually referred to as the "recoil peak", even though, as we shall see now, the origins of structures in the binary and the recoil peaks are dictated by the behavior of the initial momentum-distribution function:

Let us focus in Figs. 1.1 and 1.2 on the regime of small E_2 where q is also small. Expanding Eq. (1.12) in powers of q we find

$$\tilde{\chi}(\boldsymbol{q},\boldsymbol{k}_2) = (2\pi)^{3/2}\left\{\int \mathrm{d}^3\boldsymbol{p}\,\tilde{\psi}^*_{\boldsymbol{k}_2}(\boldsymbol{p})\,\tilde{\varphi}(-\boldsymbol{p}) - q\,\hat{\boldsymbol{q}}\cdot\int \mathrm{d}^3\boldsymbol{p}\,\tilde{\psi}^*_{\boldsymbol{k}_2}(\boldsymbol{p})\,\boldsymbol{\nabla}_{\boldsymbol{p}}\tilde{\varphi}(-\boldsymbol{p}) + \cdots\right\} \tag{1.13}$$

The first term vanishes due to orthogonality. The second "dipole term" reveals that for small q the interaction of the incoming electron with the target is related to the photoemission of the initially bound state φ by a *linearly* polarized photon, with the polarization vector being along $\hat{\boldsymbol{q}}$ (for this reason the small q case is called the optical limit). Since $\hat{\boldsymbol{q}}$ enters bilinearly in the cross section the operation $\hat{\boldsymbol{q}} \leftrightarrows -\hat{\boldsymbol{q}}$ leaves the cross section invariant in the optical limit and hence the double-peak structure occurs in Fig. 1.2a. This in turn means that for small q the recoil peak has the same origin as the binary peak[3]. With increasing E_2 the momentum

[3]In the dipole approximation the angular distribution of photoelectrons emitted within the solid angle Ω by plane-polarized radiation with energy E from an isotropic (atomic) target is given by

$$\sigma_j(\Omega) = \frac{\sigma_j(E)}{4\pi}\left[1+\beta_\gamma P_2(\cos(\theta))\right] \tag{1.14}$$

where θ is the angle between the polarization direction and the photoelectron emission direction and $P_2(x)$ is the second Legendre polynomial. $\sigma_j(E)$ is the photoionization cross section into the ionic state j and β_γ is the (energy-dependent) asymmetry parameter [7–11]. Hence, the shape of the angular distribution is governed by the asymmetry parameter β_γ. For example, for a one-electron atom which resides initially in an angular momentum state l, β_γ is given by

$$\beta_\gamma = \frac{l(l-1)d_{l-1}^2 + (l+1)(l+2)d_{l+1}^2 - 6l(l+1)d_{l+1}d_{l-1}\cos(\delta_{l+1}-\delta_{l-1})}{(2l+1)\left[ld_{l-1}^2 + (l+1)d_{l+1}^2\right]} \tag{1.15}$$

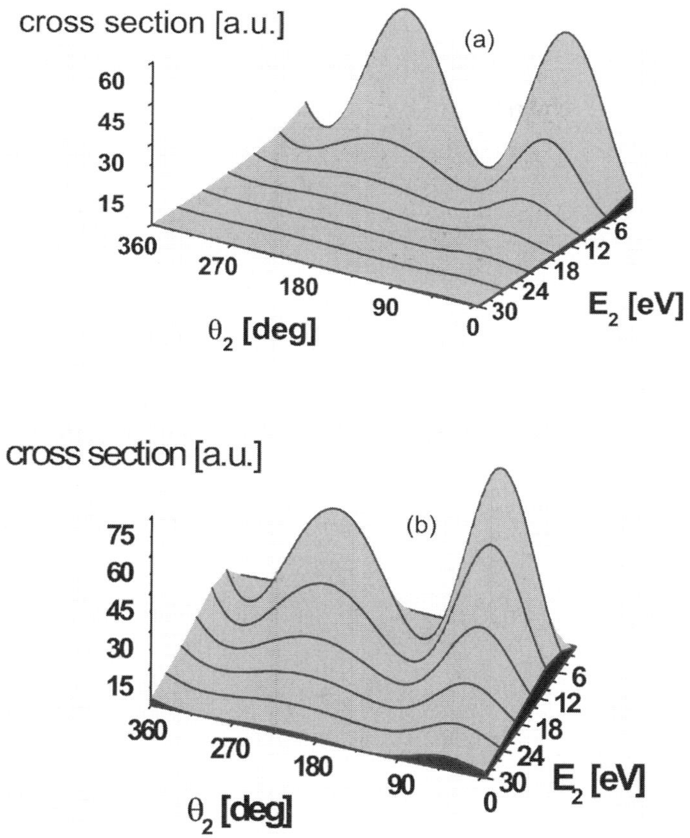

Figure 1.2: The same scattering geometry as in Fig. 1.1. In (a) the cross section is calculated within the first Born approximation whereas in (b) the final state is described by plane waves orthogonalized to the initial state.

transfer has to be increased and the directional character of the electron–electron interaction is then manifested as a break in the $\hat{q} \leftrightarrows -\hat{q}$ symmetry of the cross section, as clearly observed in Fig. 1.2a.

In the optical limit there might be some confusion about the origin of the recoil peak. Because this peak does not appear when neglecting final-state interactions, i.e., when using a plane-wave final state (cf. Fig. 1.1), it is tempting to assign the recoil peak to final state interactions. It should be pointed out however that the reason why the recoil peak in Fig. 1.2a

Here $d_{l\pm1}$ stand for the real-valued radial transition dipole matrix elements to the continuum state of the angular momentum $l \pm 1$. The cosine contains the difference of the phase shifts $\delta_{l\pm1}$ in the $l \pm 1$ partial waves.

does not show up in a plane-wave calculation is the following: From Eq. (1.13) it is clear that for small q and if the initial and the final states of the bound electron are not orthogonal the leading term is in fact the first term in the expansion, which in the case of the plane-wave approximation simplifies to $\tilde{\varphi}(\boldsymbol{k}_2)$. This deficiency of the plane-wave approximation is remedied by orthogonalizing $\psi_{\boldsymbol{k}_2}$ to φ. This we achieve by introducing the function

$$\begin{aligned} \psi^o_{\boldsymbol{k}_2}(\boldsymbol{r}_2) &= N_o\left[\psi_{\boldsymbol{k}_2}(\boldsymbol{r}_2) - \lambda\varphi(\boldsymbol{r}_2)\right], \\ \lambda &= \langle\, \varphi | \psi_{\boldsymbol{k}_2}\,\rangle = \tilde{\varphi}^*(\boldsymbol{k}_2), \\ &\Rightarrow \Psi^o_{\boldsymbol{k}_1,\boldsymbol{k}_2}(\boldsymbol{r}_1,\boldsymbol{r}_2) = (2\pi)^{-3/2}\mathrm{e}^{\mathrm{i}\boldsymbol{k}_1\cdot\boldsymbol{r}_1}\psi^o_{\boldsymbol{k}_2}(\boldsymbol{r}_2) \end{aligned} \tag{1.16}$$

where N_o is a normalization constant and $\Psi^o_{\boldsymbol{k}_1,\boldsymbol{k}_2}(\boldsymbol{r}_1,\boldsymbol{r}_2)$ is the orthogonalized two-particle final state. Using Ψ^o when calculating the cross section[4] (Fig. 1.2b) we find the same symmetry and the same behavior of the cross section as in the FBA. In particular the "recoil" peak appears at $\boldsymbol{k}_2 = -\boldsymbol{q}$. Of course final-state interactions and multiple scattering events, which are not incorporated in a plane wave treatment will also leave traces in the cross section, however the chosen scattering geometry is decisive for such structures to be observable. We shall address this point later in a more quantitative manner. Here we continue our qualitative analysis by a discussion of the generalities of the dependence of the cross section on the details of $\Psi_{\boldsymbol{k}_1\boldsymbol{k}_2}(\boldsymbol{r}_1,\boldsymbol{r}_2)$.

1.3 Approximate Formula for the Electron–Electron Ionization Cross Section

In the preceding section we discussed the behavior of the cross section when we employ the FBA or the plane-wave approximation for the final state. Some general conclusions can also be made when the final state $\Psi_{\boldsymbol{k}_1\boldsymbol{k}_2}(\boldsymbol{r}_1,\boldsymbol{r}_2)$ has a more complicated structure and the matrix elements (1.3) cannot be evaluated in a closed form. This is achieved by reformulating

[4] Using the orthogonalized plane wave final state (1.16) we infer for the cross section

$$\mathrm{d}\sigma(\boldsymbol{k}_1,\boldsymbol{k}_2) = 2\pi\frac{1}{v_0}\left|T^o\right|^2\delta(E_\mathrm{f}-E_\mathrm{i})\,\mathrm{d}^3\boldsymbol{k}_1\mathrm{d}^3\boldsymbol{k}_2, \tag{1.17}$$

$$T^o = \tilde{U}(\boldsymbol{q})\left[\tilde{\varphi}(\boldsymbol{q}-\boldsymbol{k}_2) - (2\pi)^{3/2}\,\tilde{\varphi}(\boldsymbol{k}_2)\tilde{\rho}(-\boldsymbol{q})\right] \tag{1.18}$$

where $\tilde{\rho}$ is the Fourier transform of the initial charge-density distribution $\rho(\boldsymbol{r}_2) = \varphi^*(\boldsymbol{r}_2)\varphi(\boldsymbol{r}_2)$. We note that $\lim_{q\to 0}\tilde{\rho}(-\boldsymbol{q}) \to (2\pi)^{-3/2}$ and usually $\lim_{q\gg 1}\tilde{\rho}(-\boldsymbol{q}) \to 0$. These two limiting cases can be utilized to estimate the importance of orthogonalizing the final state in a particular scattering geometry.

Eq. (1.3) in the momentum space as[5]

$$T = \int d^3\boldsymbol{p}_1 \int d^3\boldsymbol{p}_2\, \tilde{\Psi}^*_{\boldsymbol{k}_1\boldsymbol{k}_2}(\boldsymbol{p}_1,\boldsymbol{p}_2) \underbrace{\tilde{U}(\boldsymbol{k}_0-\boldsymbol{p}_1)\tilde{\varphi}^*(\boldsymbol{k}_0-\boldsymbol{p}_1-\boldsymbol{p}_2)}_{T_{\mathrm{pw}}(\boldsymbol{p}_1,\boldsymbol{p}_2)} \tag{1.21}$$

where $\tilde{\Psi}_{\boldsymbol{k}_1\boldsymbol{k}_2}(\boldsymbol{p}_1,\boldsymbol{p}_2)$ is the Fourier transform of the final state wave function and the term in the brace is the matrix element $T_{\mathrm{pw}}(\boldsymbol{p}_1,\boldsymbol{p}_2)$ evaluated within the plane-wave approximation[6]. If the mean (total) kinetic energy is larger than the mean total potential energy then $\tilde{\Psi}^*_{\boldsymbol{k}_1\boldsymbol{k}_2}(\boldsymbol{p}_1,\boldsymbol{p}_2)$ is sharply peaked for[7] $\boldsymbol{p}_1=\boldsymbol{k}_1$ and $\boldsymbol{p}_2=\boldsymbol{k}_2$. Therefore, if $T_{\mathrm{pw}}(\boldsymbol{p}_1,\boldsymbol{p}_2)$ varies smoothly with $\boldsymbol{p}_1,\boldsymbol{p}_2$ then we conclude that

$$T = T_{\mathrm{pw}}(\boldsymbol{k}_1,\boldsymbol{k}_2)\int d^3\boldsymbol{p}_1 \int d^3\boldsymbol{p}_2\, \tilde{\Psi}^*_{\boldsymbol{k}_1\boldsymbol{k}_2}(\boldsymbol{p}_1,\boldsymbol{p}_2) \tag{1.22}$$

$$= (2\pi)^3 T_{\mathrm{pw}}(\boldsymbol{k}_1,\boldsymbol{k}_2)\Psi_{\boldsymbol{k}_1\boldsymbol{k}_2}(\boldsymbol{r}_1=0,\boldsymbol{r}_2=0) \tag{1.23}$$

Here we made use of the fact that

$$\int d^3\boldsymbol{p}_1 \int d^3\boldsymbol{p}_2 \tilde{\Psi}^*_{\boldsymbol{k}_1\boldsymbol{k}_2}(\boldsymbol{p}_1,\boldsymbol{p}_2)$$

$$= \lim_{r_1'\to 0, r_2'\to 0} (2\pi)^{-3}\int d^3\boldsymbol{r}_1 \int d^3\boldsymbol{r}_2 \Psi_{\boldsymbol{k}_1\boldsymbol{k}_2}(\boldsymbol{r}_1,\boldsymbol{r}_2)\int d^3\boldsymbol{p}_1 \int d^3\boldsymbol{p}_2 e^{i\boldsymbol{p}_1\cdot(\boldsymbol{r}_1-\boldsymbol{r}_1')}e^{i\boldsymbol{p}_2\cdot(\boldsymbol{r}_2-\boldsymbol{r}_2')}$$

$$= (2\pi)^3\Psi_{\boldsymbol{k}_1\boldsymbol{k}_2}(\boldsymbol{r}_1=0,\boldsymbol{r}_2=0) \tag{1.24}$$

From Eq. (1.23) it is evident that, in the high kinetic energy regime, the behavior of the wave function at the origin is decisive. As discussed in [3], the short distance behavior of the wave function $\Psi_{\boldsymbol{k}_1\boldsymbol{k}_2}$ is dictated by the Fock expansion for $r_1 \to 0$ *and* $r_2 \to 0$, whereas at the two-body coalescence points $r_1 \to 0$ *or* $r_2 \to 0$ the wave function behavior is dictated by the Kato cusp conditions (Section 8.1 of [3]).

---

[5]This expression is valid when U depends on the interelectronic coordinate only. Generally, U contains an additional contribution describing the direct scattering of the incoming projectile from the residual ion, i.e.

$$U = 1/(\boldsymbol{r}_1-\boldsymbol{r}_2) - Z/r_1 \tag{1.19}$$

Hence, in addition to the expression (1.21) the transition matrix element receives the additional term

$$T_{eZ} = -Z\int d^3\boldsymbol{p}_1 \frac{1}{2\pi^2(\boldsymbol{k}_0-\boldsymbol{p}_1)^2}\int d^3\boldsymbol{p}_2\, \tilde{\Psi}^*_{\boldsymbol{k}_1\boldsymbol{k}_2}(\boldsymbol{p}_1,\boldsymbol{p}_2)\tilde{\varphi}^*(\boldsymbol{p}_2) \tag{1.20}$$

which vanishes if $\tilde{\Psi}^*_{\boldsymbol{k}_1\boldsymbol{k}_2}(\boldsymbol{p}_1,\boldsymbol{p}_2)$ and $\tilde{\varphi}^*(\boldsymbol{p}_2)$ are orthogonal (for all $\boldsymbol{p}_1$).

[6]In practice the momentum integration in Eq. (1.21) is limited by the momentum spectral density of the initial state, i.e., at $|\boldsymbol{k}_0-\boldsymbol{p}_1-\boldsymbol{p}_2| = k_{\mathrm{cutoff}}$ where $\tilde{\varphi}^*(k)\approx 0$ for $k > k_{\mathrm{cutoff}}$.

[7]We recall that physically $\boldsymbol{k}_{1/2}$ are the particles asymptotic wave numbers which are supposed to be determined by experiment.

From Eq. (1.23) we conclude that the normalization factor of the wave function $N_{\boldsymbol{k}_1\boldsymbol{k}_2}$ plays a key role: This follows from writing $\Psi_{\boldsymbol{k}_1\boldsymbol{k}_2}$ as $N_{\boldsymbol{k}_1\boldsymbol{k}_2}\Psi^{r}_{\boldsymbol{k}_1\boldsymbol{k}_2}(\boldsymbol{r}_1,\boldsymbol{r}_2)$, where

$$\int \mathrm{d}^3\boldsymbol{r}_1\mathrm{d}^3\boldsymbol{r}_2\Psi^{r*}_{\boldsymbol{k}_1\boldsymbol{k}_2}(\boldsymbol{r}_1,\boldsymbol{r}_2)\Psi^{r}_{\boldsymbol{k}'_1\boldsymbol{k}'_2}(\boldsymbol{r}_1,\boldsymbol{r}_2) = \delta(\boldsymbol{k}_1-\boldsymbol{k}'_1)\delta(\boldsymbol{k}_2-\boldsymbol{k}'_2)|N_{\boldsymbol{k}_1\boldsymbol{k}_2}|^{-2} \quad (1.25)$$

When $\Psi^{r}_{\boldsymbol{k}_1\boldsymbol{k}_2}(\boldsymbol{r}_1,\boldsymbol{r}_2)$ can be expressed near the origin in powers of r_1 and r_2 (the expansion coefficients may still be k dependent) we can expect Eq. (1.21) to follow the relation $T \sim (2\pi)^3 N_{\boldsymbol{k}_1,\boldsymbol{k}_2}T_{\mathrm{pw}}(\boldsymbol{k}_1,\boldsymbol{k}_2)$. Generally, within the approximation (1.23) the cross section (1.21) is cast as

$$\mathrm{d}\sigma(\boldsymbol{k}_1,\boldsymbol{k}_2) \approx (2\pi)^{10}\frac{1}{v_0}\left|T_{\mathrm{pw}}(\boldsymbol{k}_1,\boldsymbol{k}_2)\right|^2\overline{DOS}\,\mathrm{d}^3\boldsymbol{k}_1\mathrm{d}^3\boldsymbol{k}_2 \quad (1.26)$$

where

$$\overline{DOS} := |N_{\boldsymbol{k}_1,\boldsymbol{k}_2}|^2|\Psi^{r}_{\boldsymbol{k}_1\boldsymbol{k}_2}(\boldsymbol{r}_1=0,\boldsymbol{r}_2=0))|^2\delta(E_{\mathrm{f}}-E_{\mathrm{i}}) \quad (1.27)$$

Using the simplified formula (1.26) for a qualitative analysis one should bear in mind that $T_{\mathrm{pw}}(\boldsymbol{k}_1,\boldsymbol{k}_2)$ has to be evaluated with orthogonalized plane-waves (1.16), i.e., using Eq. (1.18), to avoid spurious effects such as those discussed in the context of Figs. 1.1 and 1.2. In addition we remark that Eq. (1.26) does not mean that the radial part of the wave function $\Psi_{\boldsymbol{k}_1\boldsymbol{k}_2}$ can be arbitrary; as is clear from Eq. (1.25) $N_{\boldsymbol{k}_1\boldsymbol{k}_2}$ derives from an integration over the probability density associated with $\Psi_{\boldsymbol{k}_1\boldsymbol{k}_2}$ over the whole configuration space.

To put the meaning of Eq. (1.26) in the context of established terms of many-body physics we reconsider Eq. (1.21) and write the total cross section as follows[8]

$$\begin{aligned}\sigma &= (2\pi)^4\frac{1}{v_0}\int \mathrm{d}^3\boldsymbol{r}_1\mathrm{d}^3\boldsymbol{r}_2\mathrm{d}^3\boldsymbol{r}'_1\mathrm{d}^3\boldsymbol{r}'_2\\ &\quad \varphi(\boldsymbol{r}'_2)\psi_{\boldsymbol{k}_0}(\boldsymbol{r}'_1)U(\boldsymbol{r}'_1,\boldsymbol{r}'_2)\varphi^*(\boldsymbol{r}_2)\psi^*_{\boldsymbol{k}_0}(\boldsymbol{r}_1)U^*(\boldsymbol{r}_1,\boldsymbol{r}_2)\mathcal{A}(\boldsymbol{r}_1,\boldsymbol{r}_2,\boldsymbol{r}'_1,\boldsymbol{r}'_2)\end{aligned} \quad (1.28)$$

where

$$\begin{aligned}&\mathcal{A}(\boldsymbol{r}_1,\boldsymbol{r}_2,\boldsymbol{r}'_1,\boldsymbol{r}'_2,E_{\mathrm{i}}=E_0+\epsilon_i)\\ &\quad = \int \mathrm{d}^3\boldsymbol{k}_1\mathrm{d}^3\boldsymbol{k}_2\Psi_{\boldsymbol{k}_1\boldsymbol{k}_2}(\boldsymbol{r}_1,\boldsymbol{r}_2)\Psi^*_{\boldsymbol{k}_1\boldsymbol{k}_2}(\boldsymbol{r}'_1,\boldsymbol{r}'_2)\delta(E_{\mathrm{f}}-E_{\mathrm{i}})\end{aligned} \quad (1.29)$$

The meaning of the quantity $\mathcal{A}$ is inferred from the spectral representation of the two-particle Green's function

$$G^+(\boldsymbol{r}_1,\boldsymbol{r}_2,\boldsymbol{r}'_1,\boldsymbol{r}'_2,z) = \int \mathrm{d}^3\boldsymbol{k}_1\mathrm{d}^3\boldsymbol{k}_2\frac{\Psi_{\boldsymbol{k}_1\boldsymbol{k}_2}(\boldsymbol{r}_1,\boldsymbol{r}_2)\Psi^*_{\boldsymbol{k}_1\boldsymbol{k}_2}(\boldsymbol{r}'_1,\boldsymbol{r}'_2)}{z-E_{\boldsymbol{k}_1\boldsymbol{k}_2}+\mathrm{i}0^+} \quad (1.30)$$

[8] We note the correspondence of Eq. (1.28) to the Caroli formula derived from the nonequilibrium Green's function [16]. According to this formula the cross section is the expectation value of the effective operator $M = U^\dagger\mathcal{A}U$ where $\mathcal{A}$ is the nonlocal density of states per unit energy, volume and momentum.

where $E_{\boldsymbol{k}_1\boldsymbol{k}_2}$ is the two particle energy. Hence we deduce that[9]

$$\mathcal{A}(\boldsymbol{r}_1, \boldsymbol{r}_2, \boldsymbol{r}'_1, \boldsymbol{r}'_2, E_0 + \epsilon_i) = -\frac{1}{\pi}\Im G^+(\boldsymbol{r}_1, \boldsymbol{r}_2, \boldsymbol{r}'_1, \boldsymbol{r}'_2, E_0 + \epsilon_i) \tag{1.31}$$

which is the nonlocal two-particle density of state per unit volume (in the six-dimensional space) and per unit energy evaluated at the (continuum) energy[10] $E_{\rm i}$. Eq. (1.29) evidences that

$$\begin{aligned}\mathcal{A}(\boldsymbol{r}_1 = 0, \boldsymbol{r}_2 = 0, \boldsymbol{r}'_1 = 0, \boldsymbol{r}'_2 = 0, E_{\rm i}) &= -\frac{1}{\pi}\Im G^+(0,0,0,0,E_{\rm i}), \\ &= \int \mathrm{d}^3\boldsymbol{k}_1 \mathrm{d}^3\boldsymbol{k}_2 \overline{DOS}(E_{\rm i}) \end{aligned} \tag{1.35}$$

Therefore, the physical interpretation of Eq. (1.26) is as follows: in the high kinetic energy regime the fully differential cross section is given by the plane wave scattering cross section weighted with $\overline{DOS}$, the two-particle state density per unit volume, unit energy and per unit momentum. Generally, $\overline{DOS}$ derives from the imaginary part of the retarded Green's function according to Eq. (1.35).

1.3.1 Example: An Atomic Target

An instrumental example that illustrates the use of Eqs. (1.21) and (1.23) is the case where $\Psi_{\boldsymbol{k}_1\boldsymbol{k}_2}$ describes two continuum independent electrons moving in the field of a heavy point charge Z, i.e.,

$$\begin{aligned}\Psi^{2\mathrm{C}}_{\boldsymbol{k}_1\boldsymbol{k}_2}(\boldsymbol{r}_1, \boldsymbol{r}_2) = \psi_{\boldsymbol{k}_1}(\boldsymbol{r}_1)\psi_{\boldsymbol{k}_2}(\boldsymbol{r}_2) &= (2\pi)^{-3}\exp(\mathrm{i}\,\boldsymbol{r}_1\cdot\boldsymbol{k}_1 + \mathrm{i}\,\boldsymbol{r}_2\cdot\boldsymbol{k}_2) \\ &\quad N_1\,{}_1F_1\,(\mathrm{i}\alpha_1, 1, \;-\mathrm{i}[k_1 r_1 + \boldsymbol{k}_1\cdot\boldsymbol{r}_1]\,) \\ &\quad N_2\,{}_1F_1\,(\mathrm{i}\alpha_2, 1, \;-\mathrm{i}[k_2 r_2 + \boldsymbol{k}_2\cdot\boldsymbol{r}_2]\,)\end{aligned} \tag{1.36}$$

---

[9] Note that $\lim_{\eta\to 0^+}\frac{\eta}{x^2+\eta^2} = \pi\delta(x)$.

[10] This interpretation is also evident from the definition of Eq. (1.29): The trace over the local part of $\mathcal{A}$ gives

$$\int \mathrm{d}^3\boldsymbol{r}_1\mathrm{d}^3\boldsymbol{r}_2\mathcal{A}(\boldsymbol{r}_1, \boldsymbol{r}_2, \boldsymbol{r}_1, \boldsymbol{r}_2, E_{\rm i}) = \int \mathrm{d}^3\boldsymbol{k}_1\mathrm{d}^3\boldsymbol{k}_2\delta(E_{\boldsymbol{k}_1\boldsymbol{k}_2} - E_{\rm i}) = DOS_2(E_{\rm i})$$

which is the two-particle state density per unit energy. In case $E_{\boldsymbol{k}_1\boldsymbol{k}_2} = E_{\boldsymbol{k}_1} + E_{\boldsymbol{k}_2}$ we can write

$$DOS_2(E_{\rm i}) = \int \mathrm{d}\epsilon \left\{\int \mathrm{d}^3\boldsymbol{k}_1\delta(E_{\boldsymbol{k}_1} - \epsilon)\left[\int \mathrm{d}^3\boldsymbol{k}_2\delta(E_{\boldsymbol{k}_2} + \epsilon - E_{\rm i})\right]\right\} \tag{1.32}$$

$$= \int \mathrm{d}\epsilon\ \mathrm{dos}_1(\epsilon)\ \mathrm{dos}_2(E_{\rm i} - \epsilon) \tag{1.33}$$

$$= \frac{1}{\pi^2}\int \mathrm{d}\epsilon\,\mathrm{tr}\,\Im g_1^+(\epsilon)\ \mathrm{tr}\,\Im g_2^+(E_{\rm i} - \epsilon) \tag{1.34}$$

where g_j^+ and dos_j are the single-particle Green's function and state density (per unit energy).

where $\alpha_j = -Z/v_j$ is the parameter introduced by Sommerfeld [15] and $v_j = \hbar k_j/m$ (m is the mass of the electron). ${}_1F_1(a,b,z)$ is the confluent hypergeometric function expressed in the notation of Ref. [17]. The normalization factors N_j are given by

$$N_j = \exp\left(-\frac{\pi\alpha_j}{2}\right)\Gamma(1-\mathrm{i}\alpha_j); \quad j=1,2 \tag{1.37}$$

Here $\Gamma(z)$ is the Gamma function. In what follows we refer to this approximation by the index "2C" as it consists of two Coulomb waves. Since ${}_1F_1(a,b,z=0)=1$ we find, according to Eq. (1.21), that in the high-kinetic energy limit the matrix elements are given by $T_{2\mathrm{C}} \approx N_1N_2T_{\mathrm{PW}}(\boldsymbol{k}_1,\boldsymbol{k}_2)$. The quality of this simple result can be tested, for in the special case of Eq. (1.36) the functions $\tilde{\Psi}$ which are required to evaluate Eq. (1.21) are known: For real momenta $\boldsymbol{p}$ one introduces the convergence factor[11] ζ and finds for the Fourier transform[12]

$$\begin{aligned}\tilde{\psi}_{\boldsymbol{k}}(\boldsymbol{p}) &= -\partial_\zeta\left\{\int \mathrm{d}^3\boldsymbol{r}\frac{\exp(-\zeta r)}{r}\exp[-\mathrm{i}(\boldsymbol{p}-\boldsymbol{k})\cdot\boldsymbol{r}]\;{}_1F_1(-\mathrm{i}\alpha,\;1,\;\mathrm{i}[k\,r+\boldsymbol{k}\cdot\boldsymbol{r}])\right\}_{\lim_{\zeta\to 0}} \\ &= -\partial_\zeta\left\{\frac{4\pi}{\zeta^2+P^2}\left[\frac{(\boldsymbol{P}-\boldsymbol{k})^2-(k+\mathrm{i}\zeta)^2}{\zeta^2+P^2}\right]^{\mathrm{i}\alpha}\right\}_{\lim_{\zeta\to 0}}\end{aligned} \tag{1.39}$$

where $\boldsymbol{P} = \boldsymbol{p}-\boldsymbol{k}$. Using Eq. (1.39) one can calculate the cross section according to Eq. (1.21) and compare the results to those obtained from Eq. (1.26). For the case presented in Fig. 1.2 we find that there is hardly a difference between these two approaches.

Within the 2C approximation the two electrons are independent. An approximate final state often used to account for the correlation within these two electrons is represented by the 3C wave function $\Psi^{3\mathrm{C}}_{\boldsymbol{k}_1\boldsymbol{k}_2}$ which reads [13, 14]

$$\Psi^{3\mathrm{C}}_{\boldsymbol{k}_1\boldsymbol{k}_2}(\boldsymbol{r}_1,\boldsymbol{r}_2) = N_{12}\,\Psi^{2\mathrm{C}}_{\boldsymbol{k}_1\boldsymbol{k}_2}(\boldsymbol{r}_1,\boldsymbol{r}_2)\;{}_1F_1\left(\mathrm{i}\alpha_{12},1,\;-\mathrm{i}(k_{12}r_{12}+\boldsymbol{k}_{12}\cdot\boldsymbol{r}_{12})\right) \tag{1.40}$$

where $\boldsymbol{k}_{12} = (\boldsymbol{k}_1-\boldsymbol{k}_2)/2$, $\boldsymbol{r}_{12} = \boldsymbol{r}_1-\boldsymbol{r}_2$ and $\alpha_{12} = \frac{1}{|\boldsymbol{k}_1-\boldsymbol{k}_2|}$. N_{12} is determined by Eq. (1.37). According to Eq. (1.26) the simplest approximation to account for the electronic correlation is then to modify the plane wave cross section by $|N_1N_2N_{12}|^2$, where

$$|N_j|^2 = \frac{2\pi\alpha_j}{\exp(2\pi\alpha_j)-1}, \quad j=1,2,12 \tag{1.41}$$

[11] The convergence of the integral (1.39) can be achieved without the use of the convergence factor $\mathrm{e}^{-\zeta r}$. This is done by introducing complex momenta p. The momentum-space wave functions thus obtained are valid for all p in the upper half of the complex plane. This procedure has been introduced by Dirac for one-dimensional problems [12].

[12] Note that for $\alpha_j = 0$ the wave functions $\psi_{\boldsymbol{k}_j}(\boldsymbol{r}_j)$ reduce to plane waves. This is also inferred from the momentum-space representation (1.39) by noting that

$$\delta(\boldsymbol{p}-\boldsymbol{k}) = \pi^{-2}\lim_{\zeta\to 0}\frac{\zeta}{[\zeta^2+(\boldsymbol{p}-\boldsymbol{k})^2]^2}. \tag{1.38}$$

It should be stressed however that the reliability of this approximation rests on the validity of the peaking approximation in Eq. (1.24), i.e., the mean kinetic energy has to be larger than the mean potential energy, and consequently the relative momenta have to be large for Eq. (1.24) to be viable. For the case shown in Fig. 1.2 the calculations employing the 3C model and utilizing Eq. (1.26) yield results that are only slightly different from those presented in Fig. 1.2. The reason for this behavior is that at high energies E_j the parameters α_j tend to zero and $|N_j| \to 1$ as follows from Eq. (1.41). At lower energies the momentum-dependent functions $|N_j|$ modify the cross section in a qualitative manner, in particular $|N_{12}|$ tends exponentially to zero with diminishing magnitude of the electron–electron relative momentum $|\boldsymbol{k}_1 - \boldsymbol{k}_2|$ (cf. Eq. (1.41)).

1.3.2 Electron–Electron Cross Section for Scattering from Condensed Matter

Generally, in condensed matter the initial state cannot be specified by a single-particle state φ, e.g., for valence-band electron emission from solids induced by electron-impact a band of electronic states are generally accessible. Let these states φ_ν be characterized by the collective quantum number ν (e.g., ν may stand for the single-particle energies, the crystal momentum, the band index and the spin (cf. Appendix A)). The cross section given by Eq. (1.28) reads then

$$\sigma = (2\pi)^4 \frac{1}{v_0} \int \mathrm{d}^3\boldsymbol{r}_1 \mathrm{d}^3\boldsymbol{r}_2 \mathrm{d}^3\boldsymbol{r}_1' \mathrm{d}^3\boldsymbol{r}_2' \int \mathrm{d}\epsilon \Bigg\{ \Bigg[\sum_\nu \varphi_\nu(\boldsymbol{r}_2')\varphi_\nu^*(\boldsymbol{r}_2)\delta(\epsilon_\nu - \epsilon) \Bigg]$$
$$\psi_{\boldsymbol{k}_0}(\boldsymbol{r}_1')U(\boldsymbol{r}_1', \boldsymbol{r}_2')\psi_{\boldsymbol{k}_0}^*(\boldsymbol{r}_1)U^*(\boldsymbol{r}_1, \boldsymbol{r}_2)\, \mathcal{A}(\boldsymbol{r}_1, \boldsymbol{r}_2, \boldsymbol{r}_1', \boldsymbol{r}_2', E_0 + \epsilon) \Bigg\}$$

The expression in the square brackets is nothing but the sample's nonlocal single-particle state density per unit energy and unit volume, i.e.,

$$\sum_\nu \varphi_\nu(\boldsymbol{r}_2')\varphi_\nu^*(\boldsymbol{r}_2)\delta(\epsilon_\nu - \epsilon) = \frac{1}{\pi}\Im g^-(\boldsymbol{r}_1, \boldsymbol{r}_1', \epsilon_\nu - \epsilon) \tag{1.42}$$

where g^- is the retarded single-particle Green's function of the specimen. With this equation the cross section can be expressed generally as

$$\sigma = -\frac{16\pi^2}{v_0} \int \mathrm{d}^3\boldsymbol{r}_1 \mathrm{d}^3\boldsymbol{r}_2 \mathrm{d}^3\boldsymbol{r}'_1 \mathrm{d}^3\boldsymbol{r}'_2 \Big[\psi_{\boldsymbol{k}_0}(\boldsymbol{r}'_1) U(\boldsymbol{r}'_1, \boldsymbol{r}'_2) \psi^*_{\boldsymbol{k}_0}(\boldsymbol{r}_1) U^*(\boldsymbol{r}_1, \boldsymbol{r}_2) \int \mathrm{d}\epsilon \Im g^-(\boldsymbol{r}_1, \boldsymbol{r}'_1, \epsilon)\, \Im G^+(\boldsymbol{r}_1, \boldsymbol{r}_2, \boldsymbol{r}'_1, \boldsymbol{r}'_2, E_0 + \epsilon)\Big] \quad (1.43)$$

From this formula a hierarchy of approximations derives. For example, if the off-diagonal elements of the Green's functions are small and the density per unit volume varies smoothly in space on the scale of the variations of $\psi_{\boldsymbol{k}_0}(\boldsymbol{r}'_1)U(\boldsymbol{r}'_1, \boldsymbol{r}'_2)$ we can make the approximation $\Im g^-(\boldsymbol{r}_1, \boldsymbol{r}'_1, \epsilon_\nu - \epsilon) \approx \Im g^-(\boldsymbol{r}_1, \boldsymbol{r}_1, \epsilon_\nu - \epsilon) \approx \mathrm{dos}(\epsilon_\nu - \epsilon)/V$ where V is the space volume and $\mathrm{dos}(x)$ is the density of state per energy unit. Making a similar approximation for $\Im G^+$ we arrive at the conclusion that the cross section is proportional to a convolution of the bound-initial and final-channel density of states, an approximation often used in spectroscopic studies in condensed matter such as single photoemission [18–20].

1.3.3 Electron Scattering Cross Section from Ordered Materials

For lattice periodic systems it is advantageous to express the electronic states $\varphi_{\boldsymbol{k}}$ in momentum space (cf. Appendix A), where $\boldsymbol{k}$ is the Bloch vector (or the crystal momentum) and the energy associated with this state is $\epsilon(\boldsymbol{k})$. The cross section in a momentum space representation reads

$$\begin{aligned} \sigma &= -\frac{2}{v_0} \int \mathrm{d}^3\boldsymbol{p}_1 \mathrm{d}^3\boldsymbol{p}_2 \mathrm{d}^3\boldsymbol{p}'_1 \mathrm{d}^3\boldsymbol{p}'_2 \Big[\tilde{U}(\boldsymbol{k}_0 - \boldsymbol{p}'_1) \tilde{U}^*(\boldsymbol{k}_0 - \boldsymbol{p}_1) \\ &\qquad \tilde{\varphi}^*_{\boldsymbol{k}}(\boldsymbol{p}_1 + \boldsymbol{p}_2 - \boldsymbol{k}_0) \tilde{\varphi}_{\boldsymbol{k}}(\boldsymbol{p}'_1 + \boldsymbol{p}'_2 - \boldsymbol{k}_0) \\ &\qquad \Im G^+(\boldsymbol{p}_1, \boldsymbol{p}_2, \boldsymbol{p}'_1, \boldsymbol{p}'_2, E_0 + \epsilon(\boldsymbol{k}))\Big] \\ &= -\frac{16\pi^3}{v_0} \int \mathrm{d}^3\boldsymbol{p}_1 \mathrm{d}^3\boldsymbol{p}_2 \mathrm{d}^3\boldsymbol{p}'_1 \mathrm{d}^3\boldsymbol{p}'_2 \Big[\sum_{\boldsymbol{g}, \boldsymbol{g}'} \tilde{U}(\boldsymbol{k}_0 - \boldsymbol{p}'_1) \tilde{U}^*(\boldsymbol{k}_0 - \boldsymbol{p}_1) \\ &\qquad c^*_{\boldsymbol{g}}(\boldsymbol{k}) \delta_{\boldsymbol{g}+\boldsymbol{k}, \boldsymbol{p}_1 + \boldsymbol{p}_2 - \boldsymbol{k}_0} c_{\boldsymbol{g}'}(\boldsymbol{k}) \delta_{\boldsymbol{g}'+\boldsymbol{k}, \boldsymbol{p}'_1 + \boldsymbol{p}'_2 - \boldsymbol{k}_0} \\ &\qquad \Im G^+(\boldsymbol{p}_1, \boldsymbol{p}_2, \boldsymbol{p}'_1, \boldsymbol{p}'_2, E_0 + \epsilon(\boldsymbol{k}))\Big] \end{aligned} \quad (1.44)$$

Here $\boldsymbol{g}, \boldsymbol{g}'$ are reciprocal lattice vectors and $c_{\boldsymbol{g}}(\boldsymbol{k})$ are the expansion coefficients of the Bloch state in a plane wave basis (cf. Appendix A), i.e., $c_{\boldsymbol{g}}(\boldsymbol{k})$ describe the modulations of a plane wave with the wave vector $\boldsymbol{k}$ due to the crystal periodicity. Hence if the periodic potential is weak (with respect to the electron kinetic energy) the coefficients $c_{\boldsymbol{g}}(\boldsymbol{k})$ are strongly peaked at $\boldsymbol{g} = \boldsymbol{k}$ and the expression (1.44) simplifies considerably. A more detailed analysis of the structure of Eq. (1.44) and its physical significance will be given in a later chapter.

1.3.4 Initial- vs. Final-state Interactions

The above analysis is based on the so-called prior form (1.3) of the T matrix, i.e., we start from a well-prepared (asymptotic), uncorrelated initial state and let the system interact. Hence the two-particle wave function $\Psi_{\boldsymbol{k}_1\boldsymbol{k}_2}$ describes the interacting system with outgoing wave boundary conditions. Equivalently, one can operate within the post form in which case the asymptotic final state is employed and one needs then an expression for the interacting two-particle state $\Psi_{\epsilon_i,\boldsymbol{k}_0}$ with incoming wave boundary conditions. These two pictures, although equivalent in principle, in general yield different predictions because evaluation of the T matrix elements involves, in general, the use of approximate wave functions. Considering $T_{\mathrm{fi}} = \langle\, \boldsymbol{k}_1, \boldsymbol{k}_2|U|\Psi_{\epsilon_i,\boldsymbol{k}_0}\,\rangle$ and performing similar steps as those done above one concludes that the qualitative statements we made above remain unchanged regardless of the form used for the T matrix. In what follows we will use both pictures and will provide some concrete analysis of T_{fi}.

1.4 Averaged Electron–Electron Scattering Probabilities

For a variety of purposes integrated cross sections are required, e.g., the electron–electron relaxation time in solids is generally determined by the integral cross section. Much research has been devoted to the description of integral cross section. In particular the total cross section $\sigma(E_0)$ near the escape threshold has been studied in considerable detail, see for example [21–30] and references therein. The purpose here is not to repeat the conclusions of these studies but rather to look at the integrated cross section from a rather general qualitative viewpoint.

From Eq. (1.21) and (1.7) it is evident that the integrated cross sections are determined by three factors. 1. The form factor of the scattering potential, 2. the initial, and 3. the final momentum-distribution functions. There are a variety of situations where the influence of some of these factors on the cross section is predictable analytically, e.g. for tightly bound orbitals the quantity $\tilde{\varphi}$ varies smoothly in momentum space; furthermore, for strongly screened electron–electron interaction, $\tilde{U}$ assumes a particularly simple form. The question we pose here is what is the generic structure of the cross section in these two cases.

1.4.1 Integrated Cross Section for Strongly Localized States

The general structure of the angular and the energy integrated cross section for atomic targets is deducible from the case in which the initial state is strongly localized, e.g., if $Z \gg 1$. The initial-state momentum distribution $\tilde{\varphi}(\boldsymbol{k}_{\text{rec}})$ which enters Eqs. (1.21) and (1.7) can then be regarded as a smooth function on the scale of the variation of $\tilde{U}$. Hence, in this case the cross section is determined by the final state density and by $\tilde{U}$. For a systematic analysis let us at first employ a plane wave for the final state, meaning that the form factor is the determining quantity for the cross section. In what follows we consider a Coulomb screened potential $U = \mathrm{e}^{-\lambda|\boldsymbol{r}_1-\boldsymbol{r}_2|}/|\boldsymbol{r}_1 - \boldsymbol{r}_2|$ which is the form of the electron–electron interaction in the Thomas–Fermi model of screening (cf. Appendix B for more details and a brief derivation). No screening means that the inverse screening length λ tends to zero. From Eq. (1.7) we conclude then[13]

$$\mathrm{d}\sigma(\boldsymbol{k}_1, \boldsymbol{k}_2) \approx \frac{4}{v_0} \frac{\delta(E_\mathrm{f} - E_\mathrm{i})}{(|\boldsymbol{k}_0 - \boldsymbol{k}_1|^2 + \lambda^2)^2} \mathrm{d}^3\boldsymbol{k}_1 \mathrm{d}^3\boldsymbol{k}_2 \tag{1.45}$$

Taking the z-axis to be along $\hat{\boldsymbol{q}}$ and transforming (1.45) into spherical coordinate in the space spanned by $(\boldsymbol{k}_1,\ \boldsymbol{k}_2)$ we find

$$\frac{\mathrm{d}\sigma}{\mathrm{d}\theta_1 \mathrm{d}E_1} = \frac{16\pi^2}{\sqrt{2E_0}} \frac{\sqrt{E_1}\sqrt{E_0 + \epsilon_i - E_1}}{\left(E_0 + E_1 - 2\sqrt{E_1 E_0}\cos\theta_1 + \lambda^2/2\right)^2} \tag{1.46}$$

Performing the integration over θ_1 yields the result

$$\frac{\mathrm{d}\sigma}{\mathrm{d}E_1} = \frac{32\pi^2}{\sqrt{2E_0}} \frac{\sqrt{E_1}\sqrt{E_0 + \epsilon_i - E_1}}{(E_0 - E_1)^2 + \lambda^2(\frac{\lambda^2}{4} + (E_0 + E_1))} \tag{1.47}$$

The total cross section $\sigma(E_0)$ is then given by

$$\sigma(E_0) = \int_0^{E_0+\epsilon_i} \mathrm{d}E_1 \frac{\mathrm{d}\sigma}{\mathrm{d}E_1} \tag{1.48}$$

Here θ_1 is the electron polar angle.

This simple formula gives a direct insight into the behavior of the integrated cross section: To illustrate Eq. (1.48), in Fig. 1.3 the total cross section $\sigma(E_0)$ is plotted against E_0 and λ. In accord with what is found experimentally (cf. [31] and references therein), the shape of

[13] It should be noted that the Thomas–Fermi (TF) model is a static, long-wavelength approximation for the description of the screening of the potential of a charge immersed in an electron gas. Hence, it does not capture the physics at short distances, e.g., as in close collision processes or the short-range behavior of the screening of an impurity potential. In the present context however the cross section is largest in the case of far collisions and hence the TF model provides a useful tool for a qualitative understanding.

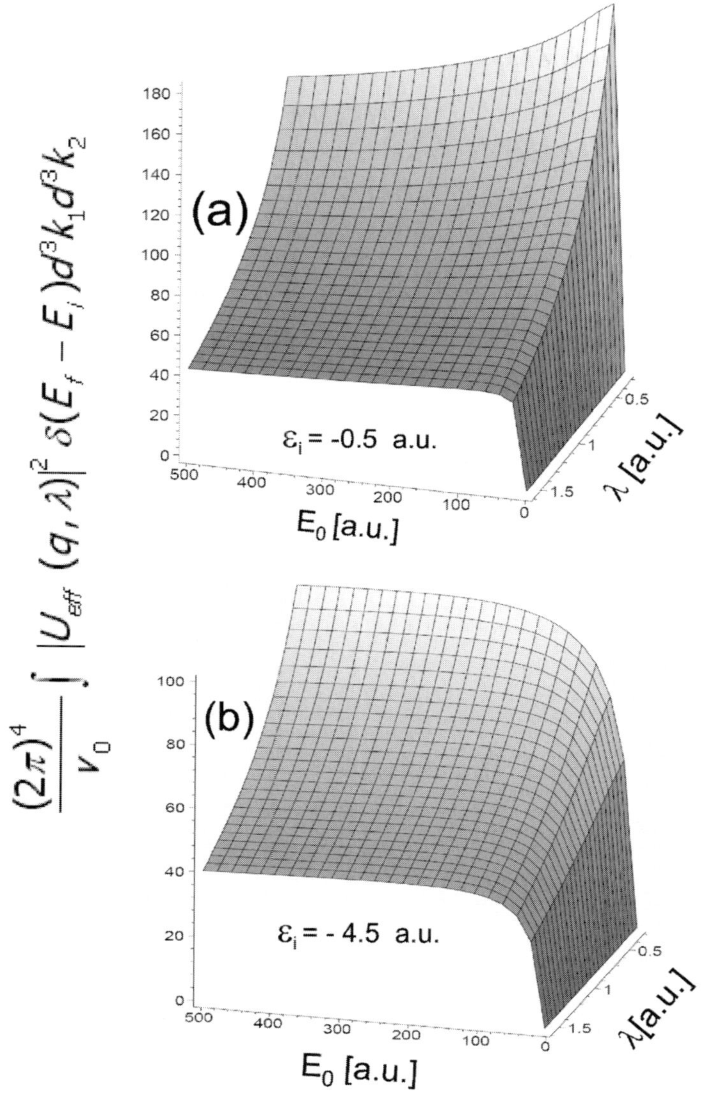

Figure 1.3: The total cross section evaluated according to the approximate formula (1.47). The dependence on the incident energy E_0 and the screening parameter λ is depicted for an initial binding energy $\epsilon = -0.5$ a.u. (inset a) and $\epsilon_i = -4.5$ a.u. (inset b).

the cross section possesses a peak structure around few times the binding energy ϵ_i (i.e., the ionization potential). The position of this peak is obtained by finding the maxima of the expression (1.48) with respect to E_0. We recall that the shape of the cross section as derived above is dictated by the form factor of the potential. For targets with a large number of delocalized electrons screening effects may well be important and these can be approximated in the long-wavelength limit by the Thomas–Fermi potential $\tilde{U}(q) \sim 1/(q^2+\lambda^2)$.

For strong screening, more precisely for $\lambda \gg q$, the volume in which scattering takes place shrinks and consequently the cross section drops, as evident from Fig. 1.3a as well as from Eqs. (1.45)–(1.48). For $\lambda \gg q$ the scattering potential is almost constant (cf. Eq. (1.45)) and is then governed by the state density per energy and momentum in the final channel. This implies a flattening of the total cross section, as observed in Fig. 1.3b (cf. also Eqs. (1.46)–(1.48)). This behavior of the cross section is supported by experiments (and more elaborate theory) on polarizable targets such as C_{60} fullerenes and large molecules [32–37].

Another noteworthy point is the dependence of the cross section on the binding energy ϵ_i. To appreciate the effect of varying ϵ_i on the cross section we express Eqs. (1.46)–(1.48) in terms of the excess energy $E_{\text{ex}} = E_0 + \epsilon_i = E_1 + E_2$. Equation (1.46) for example becomes (note the binding energy ϵ is negative)

$$\frac{\mathrm{d}\sigma}{\mathrm{d}\theta_1 \mathrm{d}E_1} = \frac{16\pi^2}{\sqrt{2E_{\text{ex}} - \epsilon_i}} \frac{\sqrt{E_1}\sqrt{E_{\text{ex}} - E_1}}{\left(E_{\text{ex}} - \epsilon_i + E_1 - 2\sqrt{E_1(E_{\text{ex}} - \epsilon_i)}\cos\theta_1 + \lambda^2/2\right)^2} \quad (1.49)$$

Hence, an increased ϵ_i has a similar effect as an increase in the inverse screening length λ, i.e. the total cross section maximum shifts towards higher E_{ex} with increasing ϵ_i and the cross section acquires a tableaux-type form, as demonstrated in Fig. 1.3b.

Now we turn to the discussion of the behavior of the angular distribution of the integrated cross section (Eq. (1.45)). The following remarks can be made:

1. Generally forward scattering is favored, i.e., the cross section is largest for $\theta_1 \approx 0$.
2. For systems with strong screening of the electron–electron interaction (i.e., for large λ and $E_{\text{ex}} \ll \lambda$) the angular distribution tends to be isotropic.
3. For $\epsilon_i \gg E_{\text{ex}}$ we also conclude from (1.49) that the angular distribution is closer to being isotropic than in the case of looser bound systems.

As for the energy distribution we conclude from Eq. (1.46) that in the low energy regime $E_{\text{ex}} \ll \epsilon_i$ the cross section behaves as $\frac{\mathrm{d}\sigma}{\mathrm{d}E_1} \sim \sqrt{E_1}\sqrt{E_{\text{ex}} - E_1}$, meaning that the cross section is symmetric with respect to $E_1 = E_{\text{ex}}/2$, vanishes at $E_1 = 0$ and $E_1 = E_{\text{ex}}$ but otherwise is smooth. In this context we recall that the density of final states which we employed in our analysis is that associated with a plane wave final state. To envision the effect of accounting for final-state interactions with the residual ion we describe the final state by the 2C wave function (Eq. 1.36) and utilize additionally Eq. (1.23). In doing so the above formula have then to be modified due to the multiplication of Eq. (1.45) by $|N_1 N_2|^2$ where N_j are given by Eq. (1.37). The angular integrated cross sections, given in Eq. (1.45) are not affected by this procedure.

To inspect the effect of final-state interactions on the energy distributions we note that $\lim_{E_j \gg 1} |N_j|^2 \to 1,\ j = 1,2$ whereas $\lim_{E_j \to 0} |N_j|^2 \to 1/\sqrt{E_j},\ j = 1,2$. This means that for atomic systems and for $E_{\text{ex}} \ll \epsilon_i$ the cross section $\frac{\mathrm{d}\sigma}{\mathrm{d}E_1}$ is finite at $E_1 = 0$ and $E_1 = E_{\text{ex}}$.

1.4.2 Low-energy Regime

The discussion of the results presented in Fig. 1.3 is based solely on the properties of the form factor of the potential, and we outlined above when this limitation of the discussion is reasonable. The other extreme limit occurs near the threshold. According to the formula (1.11), the FBA cross section involves the form factor $\tilde{U}(\boldsymbol{k}_0 - \boldsymbol{k}_1)$ where the parameter $\boldsymbol{k}_1$ is subject to the energy and momentum conservation law, meaning that k_1 varies between zero and $\sqrt{2(E_0 - \epsilon_i)}$. Near the threshold however, $E_0 \to \epsilon_i$ and hence in Eq. (1.11) we can approximate $\tilde{U}(\boldsymbol{k}_0 - \boldsymbol{k}_1) \approx \tilde{U}(\boldsymbol{k}_0)$. This means that the cross section is then determined by

$$\mathrm{d}\sigma(\boldsymbol{k}_1, \boldsymbol{k}_2)\Big|_{E_0 \to \epsilon_i} \approx 2\pi \frac{1}{v_0} \left|\tilde{U}(\boldsymbol{k}_0)\right|^2 \left|\tilde{\chi}(\boldsymbol{q}, \boldsymbol{k}_2)\right|^2 \delta(E_{\mathrm{f}} - E_{\mathrm{i}})\, \mathrm{d}^3\boldsymbol{k}_1 \mathrm{d}^3\boldsymbol{k}_2 \tag{1.50}$$

Hence near the threshold the cross section is determined by a convolution of the initial state with the continuum.

1.5 Electron–Electron Scattering in an Extended System

In contrast to the preceding section we consider now an initial state which is completely delocalized in a large volume (large with respect to the typical wavelength of the electron wave). This case is realized for example for the states of the conduction band electrons in an s–p bonded metal. Specifically, we explore qualitatively the electron–electron scattering rate in a Fermi gas for energies in the vicinity of the Fermi energy E_{F} (cf. Fig. 1.4). The initial states of the two electrons $|\boldsymbol{k}_0\rangle$ and $|\varphi_{\boldsymbol{k}}\rangle$ are specified by the wave vectors $\boldsymbol{k}_0$ and $\boldsymbol{k}$. Two

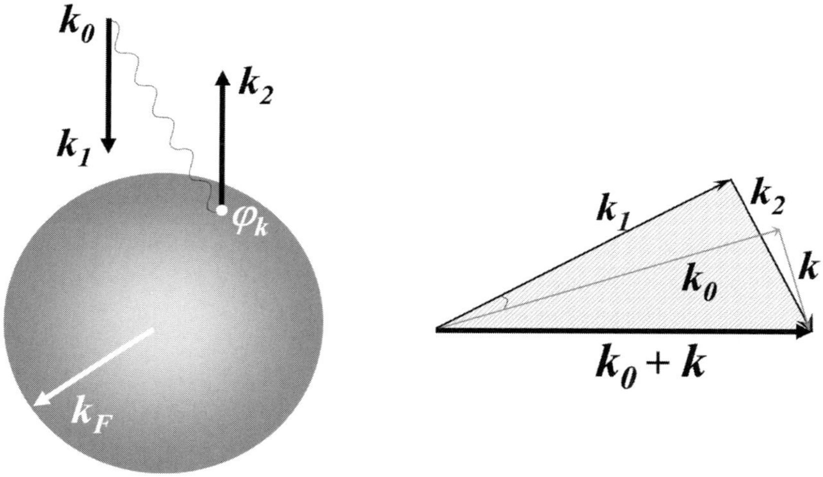

Figure 1.4: Visualization of the electron–electron scattering events near a Fermi surface with a Fermi wave vector k_{F}. A hot electron with $k_0 > k_{\mathrm{F}}$ scatters from a state φ_k with $k < k_{\mathrm{F}}$. $\boldsymbol{k}_1$ and $\boldsymbol{k}_2$ are the wave vectors of the two electrons in the final state. The drawing on the right hand side shows possible wave vectors allowed by the momentum conservations laws.

electrons inside the Fermi sphere (FS) cannot scatter from each other due to the energy conservation (1.4) combined with the Pauli principle (which allows transition to outside FS only, i.e., $E_1 + E_2 \geq 2E_{\mathrm{F}}$). Hence, the relevant situation is when k_0 is larger than the Fermi momentum k_{F} and $k \leq k_{\mathrm{F}}$ such that $E_0 + \epsilon_i \geq 2E_{\mathrm{F}}$. Here we consider the case where both k_0 and k are close to k_{F}. This allows us to obtain qualitative, but meaningful insight into the effect of the electron–electron interaction and to estimate the influence of this mechanism on the transport properties. The case where $k_0 \gg k_{\mathrm{F}}$ (such that all $|\varphi_{\boldsymbol{k}}\rangle$ are accessible) is much more involved and will be the focus of forthcoming chapters.

The scattering rate for the electron with the momentum $\boldsymbol{k}_0$ derives from the relation (cf. Eq. (1.1) while noting that $k_{\mathrm{rec}} \equiv 0$)

$$\begin{aligned}\frac{1}{\tau(E_0)} &= \frac{2\pi}{\hbar}\int \mathrm{d}^3\boldsymbol{k}\mathrm{d}^3\boldsymbol{k}_1\mathrm{d}^3\boldsymbol{k}_2\,|T|^2\,\delta(E_{\mathrm{f}}-E_{\mathrm{i}})\delta(\boldsymbol{P}_{\mathrm{f}}-\boldsymbol{P}_{\mathrm{i}}),\\ &= \frac{2\pi}{\hbar}\int \mathrm{d}^3\boldsymbol{k}\mathrm{d}^3\boldsymbol{k}_1\,|T|^2\,\delta(E_{\mathrm{f}}-E_{\mathrm{i}}) \qquad (1.51)\end{aligned}$$

Here T is given by Eq. (1.3). For a qualitative estimate of (1.51) one of two conditions is required[14]: The electron–electron interaction potential U (that enters Eq. (1.3)) is strongly

[14] In general both of these assumptions might be violated.

screened such that it can be replaced by a contact interaction with some unspecified constant scattering length (in that case $T \propto \delta(\boldsymbol{k}_0 + \boldsymbol{k} - \boldsymbol{k}_1 - \boldsymbol{k}_2)$). Alternatively, we consider the situation where $|T(\boldsymbol{k}, \boldsymbol{k}_1)|^2$ hardly varies near E_{F} such that $|T(\boldsymbol{k}, \boldsymbol{k}_1)|^2 \approx \bar{T}$ for $0 < E_1 - E_{\mathrm{F}} \ll 1$, $0 < E_{\mathrm{F}} - \epsilon_i \ll 1$. Under these circumstances the cross section is governed solely by the momentum and the energy conservation law which set strict constraints on the allowed configurations for the directions and magnitudes of the electron momenta (cf. Fig. 1.4 for the allowed configurations). Mathematically we express the scattering rate as[15]

$$\frac{1}{\tau(E_0)} \propto \frac{2\pi}{\hbar}\bar{T}\int \mathrm{d}^3\boldsymbol{k}\mathrm{d}^3\boldsymbol{k}_1\delta(E_{\mathrm{f}} - E_{\mathrm{i}}) \propto E_{\mathrm{F}}^2 \int \mathrm{d}k\mathrm{d}k_1\bigg|_{E_1+E_2=E_0+\epsilon,\boldsymbol{k}+\boldsymbol{k}_0=\boldsymbol{k}_1+\boldsymbol{k}_2} \tag{1.52}$$

To estimate the value of the integral in this equation subject to the restrictions set by the Pauli principle we note the following. Since we have chosen $k_0 \approx k_{\mathrm{F}}$, all the momenta are comparable in magnitude and are close to k_{F}, i.e., $k_1 \approx k_0 + k - k_2$ where

$$k_0, k_1, k_2 \geq k_{\mathrm{F}}, \quad k \leq k_{\mathrm{F}} \tag{1.53}$$

Hence $k_1 < k_0 + k - k_{\mathrm{F}}$ which means (cf. Eq. (1.53)) that $k_{\mathrm{F}} < k_0 + k - k_{\mathrm{F}} \Rightarrow k > 2k_{\mathrm{F}} - k_0$. We conclude therefore that the intervals within which k and k_1 vary are given, respectively, by $0 > k - k_{\mathrm{F}} > k_{\mathrm{F}} - k_0$ and $0 < k_1 - k_{\mathrm{F}} < (k_0 - k_{\mathrm{F}}) + (k - k_{\mathrm{F}})$. These inequalities set now the integration limits (due to the Pauli principle) for (1.52). Performing the integral we find the result

$$\int_{k=k_{\mathrm{F}}-k_0}^{0} \mathrm{d}k \int_0^{k_1=k_0+k-2k_{\mathrm{F}}} \mathrm{d}k_1 = \frac{(k_0 - k_{\mathrm{F}})^2}{2} \tag{1.54}$$

Inspecting Eqs. (1.52) and (1.54) we conclude that the cross section for the scattering of a hot electron with energy E_0 is proportional to $E_{\mathrm{F}}^2(k_0 - k_{\mathrm{F}})^2 \propto E_{\mathrm{F}}(E_0 - E_{\mathrm{F}})^2 = E_{\mathrm{F}}(\delta E_0)^2$. Here we introduced the energy deviation from E_{F} as $\delta E_0 = E_0 - E_{\mathrm{F}}$. The scattering rate $1/\tau$ behaves then as δE_0^2 (more precisely the collision time τ scales as $\hbar E_{\mathrm{F}}/(\delta E_0^2)$ which has a time dimension). We recall that Eq. (1.52), i.e., $E_{\mathrm{F}}(\delta E_0)^2$ is a measure for the available phase space volume. On the other hand the total volume of the FS is $(k_{\mathrm{F}}^3)^2 \propto E_{\mathrm{F}}^3$. Therefore the number of electrons that participate in the electron–electron scattering can be estimated to be $(\delta E_0/E_{\mathrm{F}})^2$ which diminishes when $E_0 \to E_{\mathrm{F}}$. For example, let us consider a thermally excited electron. At room temperature $\delta E_0 \approx k_{\mathrm{B}}T \approx 1/40$ eV, where T is the temperature and k_{B} is the Boltzmann constant. If $E_{\mathrm{F}} = 2.5$ eV we conclude that only 10^{-4} electrons

[15] Note that the momentum conservation law determines $\boldsymbol{k}_2$ and the energy conservation law determines the angle between $\boldsymbol{k}_2$ and $\boldsymbol{k}_1$. In addition k_1 and k are close to k_{F}.

participate in scattering with this electron. In general the typical value for the excited (quasi-particles) energy δE_0 is $k_\mathrm{B}T \ll E_\mathrm{F}$ and therefore we infer that $\frac{\hbar}{(E_0-E_\mathrm{F})\tau} \approx \frac{E_0-E_\mathrm{F}}{\epsilon_\mathrm{F}} \ll 1$. Even though the electron–electron interaction energy is of the order of E_F, the Fermi surface is stabilized because of the Pauli principle. Effects of electron–electron interactions can be treated to a leading order as a re-normalization of the particle mass and a relatively weak damping with the rate τ^{-1}. This means, in the presence of the electron–electron interaction the particle distribution function still has a relatively sharp jump at k_F (which defines the FS). The jump magnitude is related to the wave function overlap between a free electron and a quasiparticle at the FS (and also to the effective mass of the quasiparticle). The above argumentation underlies the Landau theory of Fermi liquids [38] which states that a strongly interacting electron gas can be mapped onto a weakly interacting gas of quasiparticles (qp), i.e., electrons 'dressed' by other electrons; the term qp indicates the fact that qp have a finite life time determined by the collision time. This means the qp peak in the spectral function has a finite width when the qp energy approaches the FS.

The qp picture proved valuable for the understanding of a variety of material properties; in many situations however it is not applicable. For example in one-dimensional systems near the Fermi points the energy spectrum is nearly linear for a single branch linear dispersion. Hence, the energy and the momentum conservation laws are in fact the same and they impose therefore less restrictions than in the three-dimensional case. Indeed perturbative corrections to account for even weak electron–electron interactions in 1D systems turned out to be divergent.

2 Spin-effects on the Correlated Two-electron Continuum

The spin-dependent scattering of electrons in matter has been exploited in a variety of important technological applications, such as the giant magneto resistance and magnetic sensors [39]. To explore further potential applications and to understand how polarized electrons propagate and excite (spin-polarized) electronic systems, various experimental techniques have been developed. For example, the spin-polarized electron-energy loss spectroscopy (SPEELS) provides important information on the fundamental elementary excitations of magnetic systems [40–42]. Surface spin waves with short wavelength as well as the Stoner spectrum (electron–hole pair spectrum)[1] have been investigated successfully by SPEELS [40,43–46]. SPEELS is particularly suited for the study of surfaces because the penetration depth of low-energy electrons is of the order of a few atomic layers. Hence, we expect the electron-pair emission, as quantified by the cross section (1.1), to be even more surface sensitive than SPEELS, for two electrons have to escape the surface[2].

In SPEELS one detects only an escaping vacuum electron that corresponds to the energy loss $\hbar\omega = E_0 - E_1$ and the momentum transfer $\boldsymbol{q} = \boldsymbol{k}_0 - \boldsymbol{k}_1$. In this chapter we focuses on a rather complementary process to SPEELS (Fig. 2.1) in which both electrons are detected. Obviously, the detection of the two electrons is possible only if the energies of the two emitted electrons are above the vacuum level. Therefore, we can investigate with this method high-energy excitations (as compared to the energy scale for magnons and Stoner excitations). Nevertheless, the low-energy excitations may influence the two-particle spectrum, as shown in Ref. [47] for the influence of the dielectric response on the correlated two-particle emission.

[1] The Stoner excitation is a two-state process: A majority band electron with energy ϵ_i interacts with an incident polarized electron that has an opposite spin projection and an energy $E_0 > E_{\text{vac}}$ (E_{vac} denotes the vacuum level, cf. Fig. 2.2). The ground state electron with energy ϵ_i is then elevated to a vacuum (detector) state with the energy E_1. The projectile electron loses energy and ends up occupying an empty hot-electron state in the minority band with an energy E_2 (usually $E_2 < E_{\text{vac}}$).

[2] This advantage comes at the expense of much lower counting rates as compared to SPEELS.

Electronic Correlation Mapping: From Finite to Extended Systems. Jamal Berakdar

ISBN: 3-527-40350-7

This issue is however still largely unexplored. On the other hand, as is intuitively clear from Fig. 2.1, the spin-resolved detection of the two electrons offers a way to explore the influence of the spin on the angular and energy-resolved electron–electron scattering.

For atomic targets a considerable body of work exists on the influence of exchange of the two electrons and/or spin–orbit interactions on electron ionizing collisions [48]. For example, the spin–orbit interaction modifies strongly the spin-polarized electron-impact ionization cross section from the K and L shells of heavy-metal targets [49–51]. The influence of exchange on fully resolved cross sections for the electron-impact ionization of light atoms has also been investigated [52, 53], some of the results of these experiments will be discussed in this work. Effects arising from exchange and spin–orbit coupling may well be present simultaneously, as occurs in the case of the polarized-electron scattering from heavy targets whilst the fine structure of the final-ion state is resolved [54–59].

2.1 Generalities on the Spin-resolved Two-electron Emission

To keep the discussion as general as possible we consider in this section the role of spins in the electron-pair emission from a perfectly ordered surface. The results are then specialized to the case of bulk materials and atomic targets. In forthcoming chapters we treat the process of two-electron escape from binary alloys. In contrast to Section 1.5, we note that the energies of the electrons we are considering here are quite high as compared to the Fermi energy and hence the electron–electron interaction plays a decisive role, as will be shown below.

The process under consideration is shown schematically in Fig. 2.1. The plane spanned by $\boldsymbol{k}_1$ and $\boldsymbol{k}_2$ (and containing $\boldsymbol{k}_0$) is oriented in some crystallographic direction of the sample. The polarized incoming electron beam is describable as a microcanonical ensemble by means of the density operator ρ^{s_1} with matrix elements $\rho^{s_1}_{m_{s_1} m'_{s_1}}$. As indicated in Fig. 2.1, m_{s_1} stands for the projection of the projectile electron's spin $\boldsymbol{s}_1$ along a quantization axis. This axis is usually given by the (natural) spin polarization axis of the electronic states of the sample, e.g., by the magnetization direction. We assume furthermore that $\rho^{s_1}_{m_{s_1} m'_{s_1}}$ is diagonalized by an appropriate unitary transformation and operate henceforth with $\rho^{s_1}_{m_{s_1} m_{s_1}}$. In the standard

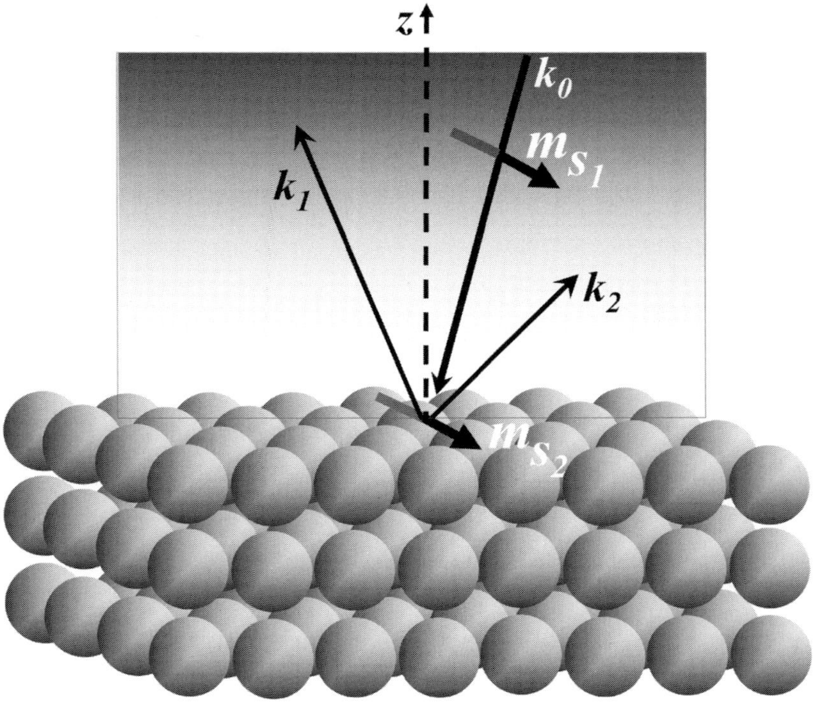

Figure 2.1: A schematic of the two-electron emission from an ordered surface upon the impact of a polarized electron with the wave vector $\boldsymbol{k}_0$ and the spin projection m_{s_1}. The momenta of the two emitted electrons $\boldsymbol{k}_1$ and $\boldsymbol{k}_2$ and $\boldsymbol{k}_0$ are in the same plane. m_{s_2} symbolizes the spin projection of the sample's electronic states.

representation ρ^{s_1} is expanded linearly in terms of the Pauli matrices $\boldsymbol{\sigma}$, i.e., one writes

$$\rho^{s_1} = \mathbf{1} + \boldsymbol{P}_0 \cdot \boldsymbol{\sigma} \tag{2.1}$$

where $\boldsymbol{P}_0$ is the spin polarization vector of the incident electron beam.

Appendix A provides a brief summary on how to describe electronic states in a periodic (crystal) potential. Here we assume these states to be spin polarized and utilize for their theoretical treatment the density matrix $\bar{\rho}^{s_2}_{m_{s_2} m_{s_2}}$ where s_2 is the spin quantum number of the sample's electronic states. m_{s_2} stands for the corresponding magnetic sublevels. As sketched in Fig. 2.1 we consider the sample as a composite of atomic layers that are indexed by l. Each of the infinitely extended atomic layers possesses a two-dimensional periodicity. The good quantum number associated with this (discrete) translational symmetry is the two-dimensional Bloch wave vector $\boldsymbol{k}_{\|}$ (see Appendix A for more details). Generally, the electronic states are fully described by the spin-resolved (Bloch) spectral function $w(\boldsymbol{k}_{\|}, l, \epsilon_i, \Uparrow)$, where $\Uparrow$ ($\Downarrow$) symbolizes the spin-projection direction of the majority (minority) band[3]. Technically $w(\boldsymbol{k}_{\|}, l, \epsilon_i, \Uparrow)$ derives from the trace of the imaginary part of the respective single particle Green's function of the surface [3]. The density operator $\bar{\rho}^{s_2}$ associated with the target electronic states is then given by

$$\bar{\rho}^{s_2} = w_0(\boldsymbol{k}_{\|}, l, \epsilon_i)(\mathbf{1} + \boldsymbol{P} \cdot \boldsymbol{\sigma}) \tag{2.2}$$

In this relation we denote by $w_0(\boldsymbol{k}_{\|}, l, \epsilon_i)$ the spin-averaged Bloch spectral function of the layer l. The quantity $\boldsymbol{P}$ is defined as

$$P = \frac{w(\boldsymbol{k}_{\|}, l, \epsilon_i, \Uparrow) - w(\boldsymbol{k}_{\|}, l, \epsilon_i, \Downarrow)}{w_0(\boldsymbol{k}_{\|}, l, \epsilon_i)} \tag{2.3}$$

Hence P stands for the degree of the spin polarization of the sample's electronic states.

The composite system consisting of the incoming electron and the surface is described in the noninteracting (asymptotic) regime by the density matrix ρ^S which is given by the product $\rho^S = \rho^{s_1} \otimes \bar{\rho}^{s_2}$.

In terms of density matrices the spin-dependent cross section (1.1) reads

$$\sigma(\boldsymbol{k}_2, \boldsymbol{k}_1; \boldsymbol{k}_0) = C \sum_{\substack{m_{s_1'}, m_{s_2'} \\ m_{s_1}, m_{s_2}}} \sum\!\!\!\!\!\!\int_{\alpha} T \rho^S T^\dagger \, \delta(E_\mathrm{f} - E_\mathrm{i}), \quad C = (2\pi)^4 / k_0 \tag{2.4}$$

where $m_{s_1'}, m_{s_2'}$ are the spin projections of the emitted electrons. The index α stands for all additional quantum numbers that are needed to characterize uniquely the sample and are not resolved by the experiment. The matrix elements T (cf. Eq. (1.3)) of the transition operator[4] $\mathcal{T}$ for the process depicted in Fig. 2.1 are formally written as

$$\begin{aligned} &T(\boldsymbol{k}_1, m_{s_1'}, \boldsymbol{k}_2, m_{s_2'}; \boldsymbol{k}_0, m_{s_1}, \alpha, m_{s_2}) \\ &\qquad = \langle \Psi_{\boldsymbol{k}_1, \boldsymbol{k}_2, m_{s_1'}, m_{s_2'}}(\boldsymbol{r}_1, \boldsymbol{r}_2) | \mathcal{T} | \varphi_{\epsilon_i, \alpha, s_2, m_{s_2}}(\boldsymbol{r}_2) \psi_{\boldsymbol{k}_0, s_1 m_{s_1}}(\boldsymbol{r}_1) \rangle \end{aligned} \tag{2.6}$$

[3] The band index does not appear in this notation because we assume that the energies of the incoming and emitted electrons are such that the energy conservation restricts the involved electronic states to one single band, e.g. the valence band.

[4] To give an idea about the structure of $\mathcal{T}$ we mention here that the leading order term in the electron–electron and the electron–crystal interaction, $\mathcal{T}$ has the form [67]

$$\mathcal{T} \approx U_\mathrm{surf} + U_\mathrm{ee}(\mathbf{1} + G_\mathrm{ee}^{-} U_\mathrm{surf}) \tag{2.5}$$

Here U_ee is the electron–electron interaction, G_ee^{-} is the retarded Green's function that involves the potential U_ee and U_surf is the surface scattering potential.

$\psi_{\boldsymbol{k}_0,s_1m_{s_1}}(\boldsymbol{r}_1)$ is a spinor vacuum wave describing the incident electron state. The electronic ground state of the surface is supposed to be describable by the single-particle, spin-resolved orbital $\varphi_{\epsilon_i,\alpha,s_2,m_{s_2}}(\boldsymbol{r}_2)$ with an associated single-particle binding energy ϵ_i. The indices s_2, m_{s_2} label the spin quantum numbers. The emitted electrons with spin projections $m_{s_1'}, m_{s_2'}$ are represented by the two-particle wave function $|\Psi_{\boldsymbol{k}_1,\boldsymbol{k}_2,m_{s_1'},m_{s_2'}}(\boldsymbol{r}_1,\boldsymbol{r}_2)\rangle$.

2.2 Formal Symmetry Analysis

It is instructive to analyze the symmetry properties of the cross section (2.4) without making reference to a specific approximation to $\mathcal{T}$ and without using a specific expression for the two-particle final state. This is achieved by disentangling the geometrical from the dynamical features (in the sense detailed below) of the two-electron emission process. To do that we express the density matrices (2.1) and (2.2) in terms of the statistical tensors (also called state multipoles) $\rho_{p_1q_1}$ and $\bar{\rho}_{p_2q_2}$ [60, 61],

$$\rho^{s_1}_{m_{s_1}m_{s_1}} = \sum_{p_1=0}^{2s_1} (-)^{p_1-s_1-m_{s_1}} \langle s_1 - m_{s_1}; s_1 m_{s_1} | p_1 q_1 = 0\rangle \rho_{p_1q_1=0}, \tag{2.7}$$

$$\bar{\rho}^{s_2}_{m_{s_2}m_{s_2}}(\epsilon_i,\alpha) = \sum_{p_2=0}^{2s_2} (-)^{p_2-s_2-m_{s_2}} \langle s_2 - m_{s_2}; s_2 m_{s_2} | p_2 q_2 = 0\rangle \bar{\rho}_{p_2q_2=0}(\epsilon_i,\alpha) \tag{2.8}$$

Here $\langle \cdots | \cdots \rangle$ are Clebsch–Gordon coefficients. The fact that in these equations only the components $\rho_{p_1q_1=0}$ and $\bar{\rho}_{p_2q_2=0}$ contribute is due to the assumption of the existence of a common quantization axis for the incoming electron spin and the spin of the electrons in the sample so that the density matrices are both diagonal.

The expansions (2.7) and (2.8) are reflected by a corresponding expansion of the cross section (2.4), specifically we can write

$$\sigma = \sum\!\!\!\!\!\!\!\int_{\alpha} \sum_{p_1=0}^{2s_1} \sum_{p_2=0}^{2s_2} \rho_{p_1q_1=0}\bar{\rho}_{p_2q_2=0}(\epsilon_i,\alpha) \Lambda^{p_1,p_2}_{q_1=0,q_2=0} \delta(E_\mathrm{f} - E_\mathrm{i}) \tag{2.9}$$

In this expression we introduced the dynamical functions

$$\begin{aligned} \Lambda^{p_1,p_2}_{q_1=0,q_2=0} = &\sum_{m_{s_1}} (-)^{p_1-s_1-m_{s_1}} \langle s_1 - m_{s_1}; s_1 m_{s_1} | p_1 q_1 = 0\rangle \\ &\times \sum_{m_{s_2}} (-)^{p_2-s_2-m_{s_2}} \langle s_2 - m_{s_2}; s_2 m_{s_2} | p_2 q_2 = 0\rangle \; \mathcal{F}_\alpha(m_{s_1}, m_{s_2}) \end{aligned} \tag{2.10}$$

where the functions $\mathcal{F}_\alpha(m_{s_1}, \alpha, m_{s_2})$ are given by

$$\mathcal{F}_\alpha(m_{s_1}, \alpha, m_{s_2}) = C \sum_{\substack{m_{s'_1} \\ m_{s'_2}}} T(\boldsymbol{k}_1, m_{s'_1}, \boldsymbol{k}_2, m_{s'_2}; \boldsymbol{k}_0, m_{s_1}, \alpha, m_{s_2}) \\ T^\dagger(\boldsymbol{k}_1, m_{s'_1}, \boldsymbol{k}_2, m_{s'_2}; \boldsymbol{k}_0, m_{s_1}, \alpha, m_{s_2}) \tag{2.11}$$

The complete dynamical information on the two-particle emission, for instance the influence of the scattering potentials or the role of the specific properties of the employed final state, are encompassed in $\Lambda^{p_1,p_2}_{q_1,q_2}$. In this sense the functions $\Lambda^{p_1,p_2}_{q_1,q_2}$ describe the scattering dynamics. Geometrical properties in the sense of how the initial state has been prepared or identified as naturally given are described by the state multipoles. Hence, the above transformations achieve a decoupling of the dynamics from geometry.

This re-coupling procedure is useful in that the sum over m_{s_1} (m_{s_2}) in Eq. (2.11) can be considered as the projection, i.e., the component, of a spherical tensor of rank p_1 (p_2) along the common quantization axis. $m_{s'_1}$ ($m_{s'_2}$) enters the sum in a parametrical sense. This statement follows from the fact that the m_{s_1} (m_{s_2}) dependence of T (for a given spin projections $m_{s'_1}$ and $m_{s'_2}$) is that of the magnetic sublevels of an angular momentum state, namely the m_{s_1} (m_{s_2}) dependence of the spin part of $|\psi_{\boldsymbol{k}_0,s_1m_{s_1}}(\boldsymbol{r}_1)\rangle$ ($|\varphi_{\epsilon_i,\alpha,s_2,m_{s_2}}(\boldsymbol{r}_2)\rangle$). For this reason we can view $T(\boldsymbol{k}_1, m_{s'_1}, \boldsymbol{k}_2, m_{s'_2}; \boldsymbol{k}_0, m_{s_1}, \alpha, m_{s_2})$ as the m_{s_1} (m_{s_2}) components of a spherical tensor of rank s_1 (s_2) [62,63].

The complex conjugate of T has the form

$$T^*(s_1, m_{s_1}) = (-)^{\delta - m_{s_1}} \mathcal{W}(s_1, -m_{s_1}).$$

This relation resembles formally the definition of the adjoint of a tensor operator $\mathcal{W}$. The phase δ is subject to the constraint $\delta - m_{s_1}$ is an integer [62], but otherwise can be chosen arbitrarily. Since in our case $p_1 = 0 \cdots 2s_1$ and $s_1 - m_{s_1}$ are always integer we can set $\delta - m_{s_1} = p_1 - s_1 - m_{s_1}$. The tensor product of $T(s_1, m_{s_1})$ and $T^\dagger(s_1, m_{s_1})$ (cf. Ref. [3,62]) is again a spherical tensor given by the equation

$$[T(s_1, m_{s_1}) \wedge T^\dagger(s_1, m_{s_1})]^{p_1}_{q_1=0} = \\ \sum_{m_{s_1}} (-)^{p_1 - s_1 - m_{s_1}} \langle s_1 - m_{s_1} s_1 m_{s_1} | p_1 0 \rangle \mathcal{W}(s_1, -m_{s_1}) T(s_1, m_{s_1}) \tag{2.12}$$

This conclusion together with the Eq. (2.11) makes clear that the function $\Lambda^{p_1,p_2}_{q_1=0,q_2=0}$ (for a given, fixed p_2) is the component along the common quantization axis of a spherical tensor of rank p_1. Similarly, we can view the dependence on p_1 parametrically and conclude that the functions $\Lambda^{p_1,p_2}_{q_1=0,q_2=0}$ are the components along the quantization axis of the spherical tensor with rank p_2.

2.3 Parametrization of the Spin-resolved Cross Sections

This above formal mathematical analysis can now be employed to write the cross section as additive tensorial components with well-defined transformation behavior according to their tensorial rank. For instance the tensorial component $\Lambda^{p_1=0,p_2}_{0,0}$ $(\Lambda^{p_1,p_2=0}_{0,0})$ is a *scalar* with respect to spin rotations generated by $\boldsymbol{s}_1$ $(\boldsymbol{s}_2)$. This means that the parts in the cross section containing this function represent spin averaged quantities in the $\boldsymbol{s}_1$ $(\boldsymbol{s}_2)$ spin space. Likewise, the components $\Lambda^{p_1=\mathrm{odd},p_2}_{0,0}$ $(\Lambda^{p_1,p_2=\mathrm{odd}}_{0,0})$ may be viewed as a spin *orientation* in the $\boldsymbol{s}_1$ $(\boldsymbol{s}_2)$ spin space (for $p_1 = 1$ it is a polar vector) and hence changes sign upon spin reflection, i.e.

$$\Lambda^{p_1=\mathrm{odd},p_2}_{0,q_0}(-m_{s_1}) = -\Lambda^{p_1=\mathrm{odd},p_2}_{0,0}(m_{s_1});\ \Lambda^{p_1,p_2=\mathrm{odd}}_{0,q_0}(-m_{s_2}) = -\Lambda^{p_1,p_2=\mathrm{odd}}_{0,0}(m_{s_2})$$

The quantities describing alignment effects, i.e., the deviations in the cross sections from the unpolarized case, are given by the tensorial components with even p_1 values.

In our case we have $s_1 = 1/2$ and $s_2 = 1/2$. Therefore equation (2.9) contains the following explicit terms

$$\sigma = \sum_{\alpha} \left\{ \Lambda^{0,0}_{0,0} \left[\rho_{00}\,\bar{\rho}_{00} + \rho_{00}\,\bar{\rho}_{10} \frac{\Lambda^{0,1}_{0,0}}{\Lambda^{0,0}_{0,0}} + \rho_{10}\bar{\rho}_{00} \frac{\Lambda^{1,0}_{0,0}}{\Lambda^{0,0}_{0,0}} + \rho_{10}\bar{\rho}_{10} \frac{\Lambda^{1,1}_{0,0}}{\Lambda^{0,0}_{0,0}} \right] \delta(E_{\mathrm{f}} - E_{\mathrm{i}}) \right\} \quad (2.13)$$

The interpretation of each of the terms in this expansion is as follows:

- The cross section averaged over the spin projections of the incoming electron and the spin polarization of the electronic states of the sample is given by the first term of the sum in Eq. (2.13).

- The spin asymmetry that occurs upon the inversion of the spin polarization direction of the electronic states in the case the incoming electron beam is *unpolarized* is described by the second term in Eq. (2.13).

- The third term in Eq. (2.13) is the spin asymmetry of the cross section that appears when inverting the spin polarization of the electron beam for an *unpolarized* target [65, 66].

In the absence of explicit spin interactions in the transition operator (e.g., in absence of relevant spin–orbit interactions in the sample's crystal potential) the functions $\Lambda_{0,0}^{1,0}$ and $\Lambda_{0,0}^{0,1}$ vanish.

- The last term of Eq. (2.13) involves the function $\Lambda_{0,0}^{1,1}$ which is a polar vector both in the $\boldsymbol{s}_1$ and $\boldsymbol{s}_2$ spin spaces, i.e.,

$$\begin{aligned}\Lambda_{0,0}^{1,1}(-m_{s_1}, m_{s_2}) &= -\Lambda_{0,0}^{1,1}(m_{s_1}, m_{s_2})\\ \Lambda_{0,0}^{1,1}(m_{s_1}, -m_{s_2}) &= -\Lambda_{0,0}^{1,1}(m_{s_1}, m_{s_2})\\ \Lambda_{0,0}^{1,1}(-m_{s_1}, -m_{s_2}) &= \Lambda_{0,0}^{1,1}(m_{s_1}, m_{s_2})\end{aligned} \tag{2.14}$$

From the relations (2.11) we deduce the explicit forms of $\Lambda_{0,0}^{1,1}$ and $\Lambda_{0,0}^{0,0}$ as

$$\Lambda_{0,0}^{1,1} = \frac{1}{2}\{\mathcal{F}(\downarrow,\Downarrow) + \mathcal{F}(\uparrow,\Uparrow) - \mathcal{F}(\uparrow,\Downarrow) - \mathcal{F}(\downarrow,\Uparrow)\} \tag{2.15}$$

$$\Lambda_{0,0}^{0,0} = \frac{1}{2}\{\mathcal{F}(\downarrow,\Downarrow) + \mathcal{F}(\uparrow,\Uparrow) + \mathcal{F}(\uparrow,\Downarrow) + \mathcal{F}(\downarrow,\Uparrow)\} \tag{2.16}$$

Here $\uparrow$ ($\downarrow$) symbolizes the case where the spin projection of the incoming electron is parallel (antiparallel) to the direction set by $\Uparrow$.

2.4 Exchange-induced Spin Asymmetry

The parameter $\Lambda_{0,0}^{1,1}$, as given by Eq. (2.15), quantifies, together with $\Lambda_{0,0}^{0,0}$ (Eq. (2.16)), the spin-dependent two-electron escape from polarized light targets (where "genuine" spin–flip processes do not occur), a case which will be treated further in this work. Examples of interest are the ionization of polarized atoms by polarized electrons and the electron-pair emission from an exchange-split ferromagnetic surface following the impact of spin polarized electrons.

To take advantage of the fact that the total spin of the electron pair is conserved and to clearly exhibit the role of the antisymmetry of the total wave function (Pauli principle) we rewrite $\mathcal{F}$ and T (Eq. (2.6)) in term of the total spin S and obtain (spin and spatial degrees of freedom are decoupled)

$$\mathcal{F}(m_{s_1}, m_{s_2}) = C \sum_{SM_S} |\langle s_1 m_{s_1}; s_2 m_{s_2} | S M_S\rangle|^2 \, X^{(S)}(\boldsymbol{k}_1, \boldsymbol{k}_2; \boldsymbol{k}_0, \alpha) \tag{2.17}$$

$$X^{(S)}(\boldsymbol{k}_1, \boldsymbol{k}_2; \boldsymbol{k}_0, \alpha) = \left|\left\langle \Psi_{\boldsymbol{k}_1,\boldsymbol{k}_2}^{(S)}(\boldsymbol{r}_1, \boldsymbol{r}_2)\chi_{SM_S} | T | \Phi^{(S)}(\boldsymbol{r}_1, \boldsymbol{r}_2)\chi'_{SM_S}\right\rangle\right|^2 \tag{2.18}$$

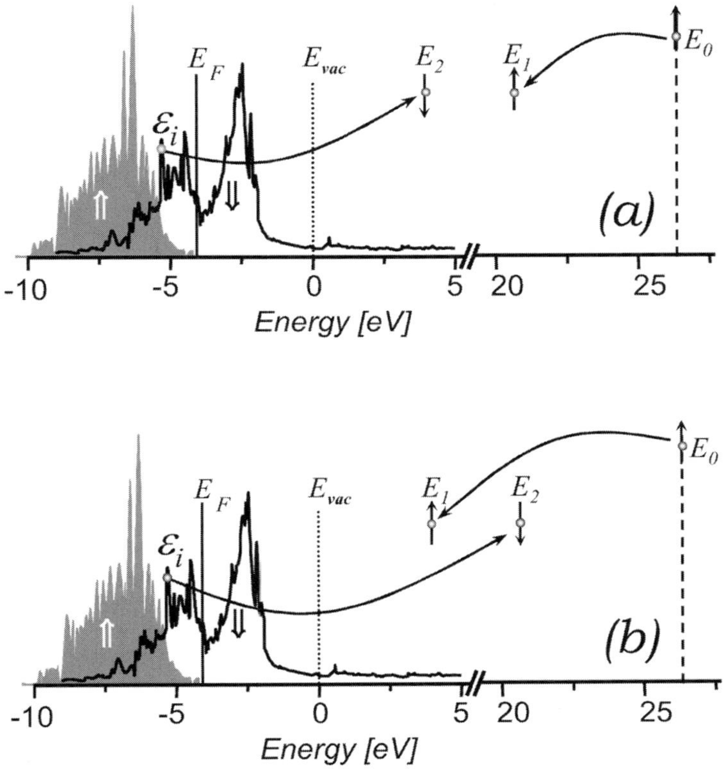

Figure 2.2: A schematic representation of the direct (a) and the exchange (b) scattering processes. E_F is the Fermi level, E_vac is the vacuum level. The density of states is that of an iron surface.

In this relation we defined the total spin-resolved cross section $X^{(S)}$ and introduced $|\chi'_{SM_S}\rangle$ as the normalized two-particle spin wave function. The spatial parts of the two-electron state in the initial and in the final channel are given by, respectively $|\Psi^{(S)}_{\boldsymbol{k}_1,\boldsymbol{k}_2}(\boldsymbol{r}_1,\boldsymbol{r}_2)\rangle$ and $|\Phi^{(S)}(\boldsymbol{r}_1,\boldsymbol{r}_2)\rangle$, i.e.,

$$|\Psi^{(S)}_{\boldsymbol{k}_1,\boldsymbol{k}_2}(1,2)\rangle = \frac{1}{\sqrt{2}}\left\{|\Psi_{\boldsymbol{k}_1,\boldsymbol{k}_2}(\boldsymbol{r}_1,\boldsymbol{r}_2)\rangle + (-)^S|\Psi_{\boldsymbol{k}_2,\boldsymbol{k}_1}(\boldsymbol{r}_1,\boldsymbol{r}_2)\rangle\right\} \tag{2.19}$$

This equation states that if the experiment is invariant under an exchange of $\boldsymbol{k}_1$ by $\boldsymbol{k}_2$ the triplet state $(S = 1)$ and the corresponding triplet transition amplitude vanish[5].

[5] We note in this context that the target may well have certain internal symmetries which are reflected in certain symmetry behavior of the two-electron final-state wave function. For example the point symmetries of the lattice structure may well be incompatible with an exchange of $\boldsymbol{k}_1$ by $\boldsymbol{k}_2$. In this case the triplet scattering cross section is not necessarily zero.

2.5 Physical Interpretation of the Exchange-induced Spin Asymmetry

The exchange-induced spin-asymmetry in the cross section is conventionally analyzed in terms of the so-called *direct* f and *exchange* g scattering amplitudes. The functions f and g are given as[6]

$$f = \langle \Psi_{\boldsymbol{k}_1,\boldsymbol{k}_2}(\boldsymbol{r}_1,\boldsymbol{r}_2)|T|\varphi_{\epsilon_i,\alpha}(\boldsymbol{r}_2)\,\psi_{\boldsymbol{k}_0}(\boldsymbol{r}_1)\,\rangle \tag{2.21}$$

$$g = \langle \Psi_{\boldsymbol{k}_2,\boldsymbol{k}_1}(\boldsymbol{r}_1,\boldsymbol{r}_2)|T|\varphi_{\epsilon_i,\alpha}(\boldsymbol{r}_2)\,\psi_{\boldsymbol{k}_0}(\boldsymbol{r}_1)\,\rangle \tag{2.22}$$

Figure 2.2(a,b) illustrates the physical meaning of the amplitudes f and g. The density of states is that of iron. In the case of an atomic target the density of state (within a single-particle description) consists of discrete points.

The direct scattering amplitude (Fig. 2.2a) describes the process in which the incoming, spin-up polarized electron with energy E_0 loses energy and ends up with the vacuum state characterized by the wave vector $E_1 < E_0$. Hereby a spin-down polarized electronic state of the target with a binding energy ϵ_i (measured with respect to the vacuum level E_{vac}) is excited to the vacuum state with the energy E_2. No spin–flip processes take place.

The exchange scattering amplitude g (Fig. 2.2b) stands for the process in which the spin-up polarized incident electron acquires in the final state the energy E_2 and the spin-down polarized excited electronic state of the target has the final-state energy E_1.

From this interpretation of g and f it is evident that the exchange-induced spin asymmetry $A^{1,1}_{0,0}$ is in fact induced by an exchange of energies, and is not due to a flip of the spins projections of the electrons. From this interpretation we conclude that if E_0 and E_1 are very large compared to E_2 (cf. Fig. 2.2), the relation applies

$$\lim_{E_0 \sim E_1 \gg E_2} |g|/|f| \to 0 \tag{2.23}$$

The validity of this expectation relies on the form factor of the (spin-independent) scattering potential which has to decay with increasing momentum transfer q. For instance, in the case of

[6] For atomic targets the residual ion in the final state usually possesses a spherical symmetry. For the wave function $\Psi_{\boldsymbol{k}_1,\boldsymbol{k}_2}(\boldsymbol{r}_1,\boldsymbol{r}_2)$ we deduce that

$$\begin{aligned}\Psi_{\boldsymbol{k}_1,\boldsymbol{k}_2}(\boldsymbol{r}_1,\boldsymbol{r}_2) &= \Psi_{-\boldsymbol{k}_1,-\boldsymbol{k}_2}(-\boldsymbol{r}_1,-\boldsymbol{r}_2)\\ \Psi_{\boldsymbol{k}_1,\boldsymbol{k}_2}(\boldsymbol{r}_1,\boldsymbol{r}_2) &= \Psi_{\boldsymbol{k}_2,\boldsymbol{k}_1}(\boldsymbol{r}_2,\boldsymbol{r}_1)\end{aligned} \tag{2.20}$$

In general these relations may well be incommensurable with the symmetry properties of the target, e.g., when the surface lacks inversion symmetry.

the electrostatic Coulomb potential (which behaves as $\sim 1/q^2$), scattering events with small momentum transfer are preferred and Eq. (2.23) applies.

The singlet ($S = 0$) and triplet ($S = 1$) scattering cross sections are expressible in terms of f and g. From Eqs. (2.19) and (2.17) we find that

$$X^{(S=0)}(\boldsymbol{k}_1, \boldsymbol{k}_2; \boldsymbol{k}_0, \alpha) = C|f + g|^2 = C|T^{(S=0)}|^2 \tag{2.24}$$

$$X^{(S=1)}(\boldsymbol{k}_1, \boldsymbol{k}_2; \boldsymbol{k}_0, \alpha) = C|f - g|^2 = C|T^{(S=1)}|^2 \tag{2.25}$$

Here we denoted the singlet (triplet) transition matrix elements by $T^{(S=0)}$ ($T^{(S=1)}$). A connection between the description in terms of the spins $\boldsymbol{s}_j$ of each of the electrons and a treatment based on the total electron spin $\boldsymbol{S}$ is established via Eq. (2.17) from which it follows that

$$\mathcal{F}(\uparrow, \Uparrow) = \mathcal{F}(\downarrow, \Downarrow) = X^{(S=1)} = C|f - g|^2 \tag{2.26}$$

$$\mathcal{F}(\downarrow, \Uparrow) = \mathcal{F}(\uparrow, \Downarrow) = \frac{1}{2}\left[X^{(S=1)} + X^{(S=0)}\right] = C|f|^2 + C|g|^2 \tag{2.27}$$

The tensorial functions $\Lambda^{1,1}_{0,0}$ and $\Lambda^{0,0}_{0,0}$ can also be written in terms of the singlet and the triplet partial cross sections, $X^{(S=0)}$ and $X^{(S=1)}$ or in terms of the direct scattering and exchange scattering amplitudes f and g, as readily inferred from Eqs. (2.15) and (2.16) that lead to the expressions

$$\Lambda^{1,1}_{0,0} = \frac{1}{2}\left[X^{(S=1)}(\boldsymbol{k}_1, \boldsymbol{k}_2; \boldsymbol{k}_0; \alpha) - X^{(S=0)}(\boldsymbol{k}_1, \boldsymbol{k}_2; \boldsymbol{k}_0, \alpha)\right] \tag{2.28}$$

$$\Lambda^{0,0}_{0,0} = \frac{1}{2}\left[3X^{(S=1)}(\boldsymbol{k}_1, \boldsymbol{k}_2; \boldsymbol{k}_0, \alpha) + X^{(S=0)}(\boldsymbol{k}_1, \boldsymbol{k}_2; \boldsymbol{k}_0, \alpha)\right] =: 2X_{\text{av}} \tag{2.29}$$

Here X_{av} is the spin averaged cross section.

2.6 Spin Asymmetry in Correlated Two-electron Emission from Surfaces

As outlined in Appendix A, the Bloch wave vector $\boldsymbol{k}$ is a good quantum number in a periodic crystal potential. For an infinitely extended surface the surface-parallel wave vector component $\boldsymbol{k}_{\parallel}$ is used to quantify the electronic states. $\boldsymbol{k}_{\parallel}$ as well as the layer l from which the emission originates are usually not determined experimentally and hence theory should average the cross section over these quantities. Expanding the initial electronic states on a basis set of a two-dimensional translational vector of the reciprocal space $\boldsymbol{g}_{\parallel}$ (as done in Eq. (A)) one concludes [67, 68] that during the emission process the surface components of the *total* wave vector of the emitted electrons

$$\boldsymbol{K}_{\|}^{+} = \boldsymbol{k}_{1\|} + \boldsymbol{k}_{2\|} \tag{2.30}$$

is conserved. This means that upon the emission process, the change of $\boldsymbol{K}_{\|}^{+}$ from its value $\boldsymbol{k}_{0\|} + \boldsymbol{k}_{\|}$ before the collision is restricted to a multiple of the surface reciprocal lattice vector $\boldsymbol{g}_{\|}$. Therefore, the sum over $\boldsymbol{k}_{\|}$ in Eq. (2.13) is in fact a summation over $\boldsymbol{g}_{\|}$, i.e.[7]

$$\begin{aligned}\sigma \propto \sum_{\boldsymbol{g}_{\|},l} \Big\{ & 2X_{\text{av}}(\boldsymbol{k}_1, \boldsymbol{k}_2; \boldsymbol{k}_0, \boldsymbol{g}_{\|}, l) \\ & \times \Big[\rho_{00}\, \bar{\rho}_{00}(\epsilon_i, \boldsymbol{\Lambda}_{\|}, l) - \rho_{10}\bar{\rho}_{10}(\epsilon_i, \boldsymbol{\Lambda}_{\|}, l) A^s(\boldsymbol{k}_1, \boldsymbol{k}_2; \boldsymbol{k}_0, \boldsymbol{g}_{\|}, l) \Big]\, \delta(E_{\text{f}} - E_{\text{i}}) \Big\}\end{aligned} \tag{2.31}$$

Here we introduced the wave vector

$$\boldsymbol{\Lambda}_{\|} = \boldsymbol{K}_{\|}^{+} - \boldsymbol{g}_{\|} - \boldsymbol{k}_{0\|} \tag{2.32}$$

The momentum dependent function A^s that enters Eq. (2.31) is the " exchange scattering asymmetry" as widely used in the literature. The explicit form of A^s in terms of $X^{(S)}$ is

$$A^s = \frac{X^{(S=0)}(\boldsymbol{k}_1, \boldsymbol{k}_2; \boldsymbol{k}_0, \boldsymbol{g}_{\|}, l) - X^{(S=1)}(\boldsymbol{k}_1, \boldsymbol{k}_2; \boldsymbol{k}_0, \boldsymbol{g}_{\|}, l)}{X^{(S=0)}(\boldsymbol{k}_1, \boldsymbol{k}_2; \boldsymbol{k}_0, \boldsymbol{g}_{\|}, l) + 3X^{(S=1)}(\boldsymbol{k}_1, \boldsymbol{k}_2; \boldsymbol{k}_0, \boldsymbol{g}_{\|}, l)} \tag{2.33}$$

For the evaluation of the terms in the sum (2.31) one requires the expressions for the state multipoles ρ_{10} and $\bar{\rho}_{10}$. By inverting the relations (2.7) and (2.8) one obtains

$$\rho_{pq} = \sum_{m_s} (-)^{p-s-m_s} \langle s - m_s; s m_s | pq \rangle \rho^{s}_{m_s m_s}. \tag{2.34}$$

This relation shows that generally all state multipoles are finite even for fully spin polarized (pure) states. For vanishing spin-dependent interactions in the initial state only the state multipoles ρ_{00}, ρ_{01}, $\bar{\rho}_{00}$ and $\bar{\rho}_{01}$ are needed to perform the sum (2.31). Equations (2.7), (2.8) and (2.34) lead us to the conclusion that

$$\rho_{00}\, \bar{\rho}_{00} = [w_0(\boldsymbol{k}_{\|}, l, \epsilon_i)]/2, \quad \rho_{10}\bar{\rho}_{10} = [w_0(\boldsymbol{k}_{\|}, l, \epsilon_i)]\, P_0 P/2.$$

The relation (2.31) is then expressible in the form

$$\sigma \propto \sum_{\boldsymbol{g}_{\|},l} w_0(\boldsymbol{\Lambda}_{\|}, l, \epsilon_i)\, X_{\text{av}}[1 + \mathcal{A}]\delta(E_{\text{f}} - E_{\text{i}}) \tag{2.35}$$

[7] Note that $\Lambda^{0,1}_{0,0}$ and $\Lambda^{1,0}_{0,0}$ vanish identically because we restricted the discussion to situations where the transition operator $\mathcal{T}$ does not contain any spin-dependent interactions (and the spin in the initial state is conserved).

The general asymmetry function $\mathcal{A}$ is evaluated according to

$$\mathcal{A} = P_0 \frac{\sum_l \left[w(\boldsymbol{A}_{\|}, l, \epsilon_i, \Downarrow) - w(\boldsymbol{A}_{\|}, l, \epsilon_i, \Uparrow)\right] \sum_{\boldsymbol{g}_{\|}} X_{\mathrm{av}} A^s \delta(E_{\mathrm{f}} - E_{\mathrm{i}})}{\sum_{l'} w_0(\boldsymbol{A}_{\|}, l', \epsilon_i) \sum_{\boldsymbol{g}'_{\|}} X_{\mathrm{av}} \delta(E_{\mathrm{f}} - E_{\mathrm{i}})} \tag{2.36}$$

$$\mathcal{A} = \frac{\sigma(\uparrow\Uparrow) - \sigma(\downarrow\Uparrow)}{\sigma(\uparrow\Uparrow) + \sigma(\downarrow\Uparrow)} \tag{2.37}$$

2.7 General Properties of the Spin Asymmetry

The preceding derivations allow several general statements concerning the properties of the exchange-induced spin asymmetry:

1. Alternatively to Eq. (2.33) we can write A^s, the spin asymmetry in the cross sections for the two-electron escape in terms of the direct $f = |f(\boldsymbol{k}_1, \boldsymbol{k}_2; \boldsymbol{k}_0, \boldsymbol{g}_{\|}, l)| \mathrm{e}^{\mathrm{i}\delta_f(\boldsymbol{k}_1, \boldsymbol{k}_2; \boldsymbol{k}_0, \boldsymbol{g}_{\|}, l)}$ and the exchange amplitudes $g = |g(\boldsymbol{k}_1, \boldsymbol{k}_2; \boldsymbol{k}_0, \boldsymbol{g}_{\|}, l)| \mathrm{e}^{\mathrm{i}\delta_g(\boldsymbol{k}_1, \boldsymbol{k}_2; \boldsymbol{k}_0, \boldsymbol{g}_{\|}, l)}$ as

$$A^s = \frac{|f|\ |g| \cos\delta}{|f|^2 + |g|^2 - |f||g| \cos\delta} \tag{2.38}$$

where $\delta = \delta_f - \delta_g$ is the relative phase between f and g. Relation (2.38) evidences that A^s is the result of a quantum interference between f and g. If A^s vanishes then $\mathcal{A}$, as given by Eq. (2.36) also vanishes. On the other hand, from Eqs. (2.38) and (2.23) we deduce

$$\lim A^s \to 0 \quad \text{if} \quad \{|f| \gg |g|, \text{ or } |f| \ll |g|, \text{ or } f \perp g, \text{ or } E_0 \sim E_1 \gg E_2\} \tag{2.39}$$

In these cases interferences between f and g are negligible.

2. From Eqs. (2.33) and (2.19) we infer that A^s varies between 1 and $-1/3$. A^s is unity if $X^{(S=1)} = 0$, which occurs for example if the experiment is invariant under an exchange of $\boldsymbol{k}_1$ and $\boldsymbol{k}_2$ (cf. Eq. (2.19) and the associated footnote).

3. For an ordered surface as a target the expression for the spin asymmetry is given by Eq. (2.36) which states that the layer-resolved A^s has to be weighted with the layer-resolved pair emission cross section X_{av}. On the other hand, X_{av} vanishes from atomic layers that are positioned (with respect to the uppermost layer) at a distance larger than the (two) electrons inelastic mean-free path. Therefore, these "deep" layers do not contribute to the expression (2.36).

4. In the case of a single layer, for instance if the target consists of a ferromagnetic layer deposited on a nonmagnetic substrate, Eq. (2.36) takes on a simpler form. In particular if the experimental arrangements are chosen such that $A^s = 1$ we deduce the important conclusion that (we note that $A^s = 0$ from the nonmagnetic layers)

$$\mathcal{A} = P_0 \frac{\left[w(\boldsymbol{\Lambda}_{\parallel}, \epsilon_i, \Downarrow) - w(\boldsymbol{\Lambda}_{\parallel}, \epsilon_i, \Uparrow)\right]}{w_0(\boldsymbol{\Lambda}_{\parallel}, \epsilon_i)} = P_0 P_{\text{film}} \tag{2.40}$$

Here P_{film} is the spin polarization of the deposited film, as defined in Eq. (2.3). Relation (2.40) is important in so far as it offers a possibility to measure the energy and wave vector resolved spin polarization of magnetic films by determining experimentally $\mathcal{A}$ and the spin-polarization of the incoming electron beam P_0.

2.7.1 Spin Asymmetry in Pair Emission from Bulk Matter

At very high electron impact energies ($E_0 > 1$ keV) [4], the fast electrons penetrate deeply into the sample and hence the break of the three-dimensional symmetry due to the presence of the surface becomes irrelevant[8]. For a three-dimensional translationally symmetric sample Eq. (2.36) simplifies to the expression

$$\mathcal{A} = P_0 P_{\text{bulk}} \frac{\sum_{\boldsymbol{g}} X_{\text{av}} A^s \delta(E_{\text{f}} - E_{\text{i}})}{\sum_{\boldsymbol{g}'} X_{\text{av}} \delta(E_{\text{f}} - E_{\text{i}})}, \tag{2.41}$$

where $\boldsymbol{g}$ is the three-dimensional reciprocal lattice vector. The spin polarization vector P_{bulk} of the bulk sample reads

$$P_{\text{bulk}} = \frac{w(\boldsymbol{\Lambda}, \epsilon_i, \Downarrow) - w(\boldsymbol{\Lambda}, \epsilon_i, \Uparrow)}{w_0(\boldsymbol{\Lambda}, \epsilon_i)}$$

In this context we note that to determine P_{bulk}, A^s should be unity. In the high energy limit the energy and the angular regions where $A^s = 1$, i.e., where triplet scattering vanishes, are expected however to be quite narrow, which may be a challenge in realizing such an experiment.

[8] Even at very high impact energies E_0 the experiment can be designed such that one emitted electron possesses a low energy E_2 and hence this electron's scattering dynamics becomes surface sensitive. However, in this case the spin asymmetry vanishes due to Eq. (2.39).

2.7.2 Spin-polarized Homogenous Electron Gas

For a three-dimensional, spin-polarized homogeneous electron gas (Stoner model) the spin polarization derives from the exchange-split density of states $\rho_{\Uparrow,\Downarrow}$. In this case Eq. (2.36) reduces to

$$\mathcal{A} = P_0 \frac{\rho_{\Downarrow} - \rho_{\Uparrow}}{\rho_{\Downarrow} + \rho_{\Uparrow}} A^s \tag{2.42}$$

Hence the exchange-induced spin asymmetry in the sample's density of state is measurable when the experiment is tuned to the configuration where $A^s = 1$, i.e., where triplet scattering is prohibited by symmetry. We remark here that, unlike the situation of an inhomogeneous electron gas as a target, for a homogenous EG, the symmetry of the final state (which depends on the chosen configuration of $\boldsymbol{k}_j$, $(j = 0, 1, 2)$ is the key factor to ensure that $T^{(S=1)} = 0$ for a particular set-up of the experiment.

2.7.3 Behavior of the Exchange-induced Spin Asymmetry in Scattering from Atomic Systems

For atomic gaseous targets the sample's polarization vector $\boldsymbol{P}$ is usually determined experimentally (the spin polarization P_a of the atomic beam) and hence Eq. (2.36) reads in this case $\mathcal{A} = P_a P_e A^s$. The quantity of interest in such a situation is

$$A^s = \frac{\sigma(S = 0, \boldsymbol{k}_1, \boldsymbol{k}_2) - \sigma(S = 1, \boldsymbol{k}_1, \boldsymbol{k}_2)}{\sigma(S = 0, \boldsymbol{k}_1, \boldsymbol{k}_2) + 3\sigma(S = 1, \boldsymbol{k}_1, \boldsymbol{k}_2)} \tag{2.43}$$

where $\sigma(S, \boldsymbol{k}_1, \boldsymbol{k}_2)$ is the cross section for the total spin S which is determined according to Eq. (1.2). So the question to be addressed here is how the scattering process, as described by the transition matrix elements is affected by the exchange of the two electrons.

For a qualitative understanding it is instructive to utilize the plane-wave approximation (PWA) for the description of the final state and then to analytically calculate the total-spin resolved cross sections (2.17) and the spin asymmetry A^s.

From Eqs. (2.24) and (2.25) and within the PWA (cf. Eq. (1.7)) we find for the singlet and the triplet transition matrix elements the following expressions

$$T^{(S=0/1)} = (2\pi)^{-3/2} \left[\tilde{U}(\boldsymbol{k}_0 - \boldsymbol{k}_1) \pm \tilde{U}(\boldsymbol{k}_0 - \boldsymbol{k}_2)\right] \tilde{\varphi}(\boldsymbol{k}_0 - \boldsymbol{K}^+) \tag{2.44}$$

From this equation we see that within the PWA and for a fixed $\boldsymbol{K}^+ = \boldsymbol{k}_1 + \boldsymbol{k}_2$ the initial state does not affect the spin asymmetry. The form factor of the potential $\tilde{U}$ determines the

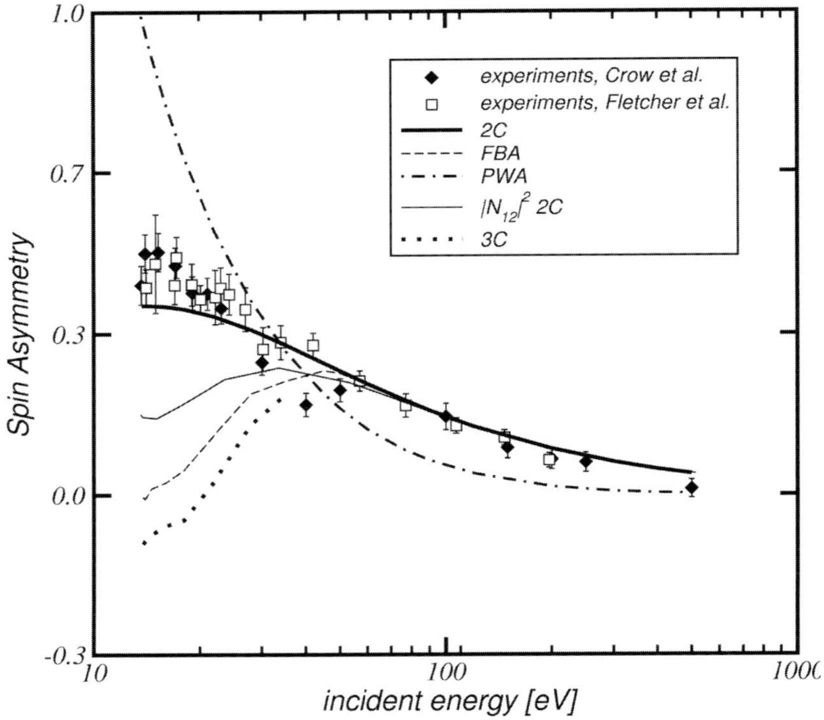

Figure 2.3: The spin asymmetry (as defined by Eq. (2.43)) in the total ionization cross section as a function of the incident-electron energy. The target is ground state atomic hydrogen. The total cross sections are obtained via a five-dimensional numerical integration over $\boldsymbol{k}_1$ and $\hat{\boldsymbol{k}}_2$. For evaluating the singlet and the triplet cross sections, as prescribed by Eq. (2.17) we employ for the spatial part of the antisymmetrized final-state wave function (2.19), the plane wave approximation (PWA), the first Born approximation (FBA), the 2C wave function (given by (1.36)), and the 3C wave function (given by (1.40)). We also show the results for the spin asymmetry when the 2C wave function is multiplied by the factor N_{12}, as given by Eq. (1.41). The corresponding curves to these approximate wave functions are indicated in the legend. The experimental data are due to Ref. [69] (full squares) and Ref. [70] (open circles).

behavior of the spin asymmetry A^s. For instance, for atomic systems the functional form of $\tilde{U}(\boldsymbol{q})$ imposes that scattering processes with small momentum transfers q are preferred. This means that in fast collisions with small q (e.g., as in Fig. 1.1) $|\tilde{U}(\boldsymbol{k}_0 - \boldsymbol{k}_1)|$ is substantially larger than $\tilde{U}(\boldsymbol{k}_0 - \boldsymbol{k}_2)$ and we find that A^s is small in this case, a trend which we already anticipated by Eq. (2.39). This high energy (and small momentum transfer) behavior of A^s is also confirmed experimentally by Fig. 2.3.

For strongly screened potentials there is a region in which the form factor becomes independent of the momentum transfer. In this case A^s also vanish, as evident from Eq. (2.44). An example of this case is the Yukawa potential which has the form factor $\sim 1/(q^2+\lambda^2)$, where λ^{-1} is the screening length. When $\lambda \gg k_0$, the form factor of this potential is then constant and A^s is zero. Here the PWA results are expected to be more reliable than for a pure (unscreened) Coulomb potential where, strictly speaking, the PWA is never valid.

2.7.4 Threshold Behavior of the Spin Asymmetry

Near threshold, i.e., for $k_0 \gg k_1$ and $k_0 \gg k_2$ we can perform a Taylor expansion of the form factor $\tilde{U}(\boldsymbol{k}_0 - \boldsymbol{k}_1)$ and find that

$$\tilde{U}(\boldsymbol{k}_0 - \boldsymbol{k}_j) = \tilde{U}(\boldsymbol{k}_0) - (\boldsymbol{k}_j \cdot \boldsymbol{\nabla}_{\boldsymbol{x}})\tilde{U}(\boldsymbol{x})\Big|_{\boldsymbol{x}=\boldsymbol{k}_0} + \cdots; \quad j = 1,2 \tag{2.45}$$

$$\begin{aligned} T^{(S=1)} &= (2\pi)^{-\frac{3}{2}} \left\{ [(\boldsymbol{k}_2 - \boldsymbol{k}_1)\cdot\boldsymbol{\nabla}_{\boldsymbol{x}}]\,\tilde{U}(\boldsymbol{x})\Big|_{\boldsymbol{x}=\boldsymbol{k}_0} \right\} \tilde{\varphi}(\boldsymbol{k}_0 - \boldsymbol{K}^+) + \mathcal{O}(k_j^2) \\ T^{(S=0)} &= (2\pi)^{-\frac{3}{2}} \left\{ 2\tilde{U}(\boldsymbol{k}_0) - [(\boldsymbol{k}_2 + \boldsymbol{k}_1)\cdot\boldsymbol{\nabla}_{\boldsymbol{x}}]\,\tilde{U}(\boldsymbol{x})\Big|_{\boldsymbol{x}=\boldsymbol{k}_0} \right\} \tilde{\varphi}(\boldsymbol{k}_0 - \boldsymbol{K}^+) + \mathcal{O}(k_j^2) \end{aligned} \tag{2.46}$$

Hence, if the form factor is smooth (and finite) near threshold, triplet scattering vanishes and the spin asymmetry tends to unity (within the PWA).

For the Coulomb potential the sums in (2.46) can be performed explicitly to all orders and we find that

$$T^{(S=1)} = \frac{1}{2\pi^2} \frac{k_2^2 - k_1^2 + 2(\boldsymbol{k}_1 - \boldsymbol{k}_2)\cdot\boldsymbol{k}_0}{|\boldsymbol{k}_0 - \boldsymbol{k}_1|^2 |\boldsymbol{k}_0 - \boldsymbol{k}_2|^2} \tilde{\varphi}(\boldsymbol{k}_0 - \boldsymbol{K}^+) \tag{2.47}$$

$$T^{(S=0)} = \frac{1}{2\pi^2} \frac{2k_0^2 + k_2^2 + k_1^2 - 2(\boldsymbol{k}_1 + \boldsymbol{k}_2)\cdot\boldsymbol{k}_0}{|\boldsymbol{k}_0 - \boldsymbol{k}_1|^2 |\boldsymbol{k}_0 - \boldsymbol{k}_2|^2} \tilde{\varphi}(\boldsymbol{k}_0 - \boldsymbol{K}^+) \tag{2.48}$$

These equations make evident that near threshold ($k_1 \to 0$, $k_2 \to 0$) the triplet scattering amplitude vanishes and hence A^s tends to unity[9].

This conclusion is not fully substantiated by the experimental data shown in Fig. 2.3, which is due to a breakdown of the PWA at low energies. Nevertheless, A^s is indeed positive near threshold and attains the PWA values in the high energy regime.

To improve on the PWA one can utilize Eq. (1.26), i.e., one accounts for the momentum-dependent density of states $\overline{DOS}(\boldsymbol{k}_1, \boldsymbol{k}_2)$. While this procedure may improve on the description of the differential cross sections, the spin-asymmetry $A^s(\boldsymbol{k}_0, \boldsymbol{k}_1, \boldsymbol{k}_2)$ in the fully resolved

[9] Note also from these equations that the triplet scattering always vanishes when $E_1 = E_2$ and $\hat{\boldsymbol{k}}_1 \cdot \hat{\boldsymbol{k}}_0 = \hat{\boldsymbol{k}}_2 \cdot \hat{\boldsymbol{k}}_0$.

cross section is a ratio of cross sections and hence does not depend on $\overline{DOS}$. Therefore, A^s provides in addition to the cross section, an independent tool to investigate the properties of the radial part of the wave function[10].

The next level in improving the final-state description is to use the FBA. Here we see a strong deviations from the experiments in the near-threshold regime (cf. Fig. 2.3). As it turned out [64], the shortcoming stems mainly from the use of an asymmetric form (Eq. (1.12)) of the radial part of the wave function and is not so much related to a deficiency (as compared to PWA results) of the $\overline{DOS}$ associated by the FBA wave function.

Improving further on the final-state wave function we utilize the 2C approximation which treats the two final state electrons in a symmetric way while the final-state electron–electron interaction is neglected. The 2C results agree remarkably well with the experiments shown in Fig. 2.3. It may appear that this finding is a mere coincidence, because the strengths of the electron–electron and the electron–nucleus interactions are of the same order and hence, a priori there is no reason to neglect one of these interactions in favor of the others. Recalling the form the transitions amplitudes within the PWA (i.e., Eqs. (2.46))[11] we see however that the value of the incident momentum k_0 plays a decisive role. On the other hand, the minimal value of k_0 is set by the initial binding energy ($k_{0,\text{min}} = \sqrt{-2\epsilon_i}$), i.e. in the initial state the electron–electron and the (bound) electron–nucleus interaction have different strengths. Increasing the charge of the residual ion increases $|\epsilon_i|$ and hence $k_{0,\text{min}}$ (while the properties of the electron–electron interaction remain unchanged). From these arguments we expect a strong dependence of A^s on the initial binding energy ϵ_i (or on the charge state Z of the residual ion).

As mentioned above, $A^s(\boldsymbol{k}_0, \boldsymbol{k}_1, \boldsymbol{k}_2)$ is not influenced by the momentum-dependent density of states $\overline{DOS}$. It should be mentioned however that A^s in the integrated cross section is modified by $\overline{DOS}$ because $\overline{DOS}$ does not cancel out when calculating the ratio (2.43) of *integrated* cross sections. An example in shown in Fig. 2.3: To account for the electron–electron interaction, at least on the level of a correlated $\overline{DOS}$, while maintaining the use of the 2C wave function we employ Eq. (1.26) and multiply the 2C (fully resolved) results by $|N_{12}|^2$, as given by Eq. (1.41), i.e., in the final state $\overline{DOS}$ is that associated with the 3C wave function. In this case we observe in Fig. 2.3 a large deviation from experiment. This indicates a defi-

[10]More precisely, statements can be made concerning the projections of the radial (final state) wave functions onto the state $U|\varphi_{\epsilon_i}\psi_{\boldsymbol{k}_0}\rangle$. Only the absolute magnitude of these projected parts is accessible experimentally (cf. Eq. (1.3)).

[11]A similar analytical analysis can be also performed within the 2C approximation.

ciency of this expression for $\overline{DOS}$ related to the 3C model. Here we stress again that $\overline{DOS}$ depend in a nonexplicit way on the overall spatial behavior of the wave function (it derives as a six-dimensional integral of the norm of the wave function over the configuration space). These arguments are endorsed by Fig. 2.3 where even larger discrepancies between the 3C calculations and the experimental data are observed, which hints at an inaccurate radial part of the 3C wave function[12]. A way to remedy, at least partially, this shortcoming of the 3C will be presented in the next chapter.

Figure 2.3 shows that near threshold A^s is almost energy independent. This behavior can be understood from an extension of the Wannier theory [21] to arbitrary L, S and π states (here L is the total angular momentum, S is the total spin and π is the parity) [71]. What is then found is that all L states and nearly all $LS\pi$-states have the same energy dependence at threshold, which explains the near-threshold behavior of A^s, however these studies do not make a statement regarding the value of A^s.

[12]This statement is more evident if we consider the fully differential cross section which is done in the next chapter. In this case A^s is independent of $\overline{DOS}$.

3 Mechanisms of Correlated Electron Emission

In this chapter we explore to what extent the qualitative aspects of the electron-pair emission that we discussed in the previous chapter remain viable when the energies and emission angles are varied arbitrarily. In particular, we will discuss the physical mechanisms that govern the behavior of the cross sections.

The early experimental realization of the energy and angular-resolved electron emission upon electron impact have been conducted at moderate incident energies, i.e., a few times the ionization potential, and for small momentum transfer q [72–76]. The two continuum electrons are detected in the same plane spanned by $\boldsymbol{k}_1$, $\boldsymbol{k}_2$ and $\boldsymbol{k}_0$ (cf. Fig. 1.1). As we discussed in the previous chapter, in this geometry the cross section is particularly large. This chapter is focused on the case of an atomic hydrogen target in its ground state, for it renders possible an interpretation of the measured cross section in terms of ionization mechanisms [75, 77–79] without having to resort to approximate initial states if many-electron targets are investigated.

In what follows we will concentrate on the understanding of the underlying physical mechanisms of the structures that are observed in the spectrum without an attempt to review all the work done in this area, in particular the impressive progress in numerical methods and approaches (see, e.g., [80–92] and further references therein) which is also paralleled with equally impressive advances in detection techniques and efficiency (see, e.g., [92, 93] and references therein) will not be discussed.

Various theoretical concepts have been put forward for the description of the process under study: For example using a perturbation approach the cross section (1.1) has been evaluated including higher orders in the Born series (beyond FBA). The calculated cross section turned out to be in quantitative agreement with the measured angular distributions [77, 94, 95]. The implementation of a (square integrable) pseudo-state close-coupling scheme (PSCC) [96, 97] leads in general to a good, although not perfect, agreement with some experimental data.

Electronic Correlation Mapping: From Finite to Extended Systems. Jamal Berakdar

ISBN: 3-527-40350-7

3.1 Exterior Complex Scaling

The exterior complex scaling (ECS) [80–83] proved reliable for a quantitative description of the ionization cross section. To outline the idea behind this method let us consider the target to be a hydrogen atom in its ground state $|\varphi\rangle$. To evaluate the matrix elements (1.3) one writes the wave function $\Psi_{\boldsymbol{k}_1\boldsymbol{k}_2}(\boldsymbol{r}_1,\boldsymbol{r}_2)$ for the total spin S as (the initial state is a symmetrized form of that given in Eq. (1.3))

$$\Psi_{\boldsymbol{k}_1\boldsymbol{k}_2,S}(\boldsymbol{r}_1,\boldsymbol{r}_2)=\frac{1}{r_1r_2}\sum_{l_1l_2LM}\psi^{LMS}_{l_1l_2}(r_1,r_2)B^{LM}_{l_1l_2}(\hat{\boldsymbol{r}}_1,\hat{\boldsymbol{r}}_2) \tag{3.1}$$

where $B^{LM}_{l_1l_2}$ is a bipolar spherical harmonics [62]. l_1 and l_2 are angular momenta of the independent electrons coupled to the total angular electron-pair momentum L. Thus, the relation $|l_1-l_2|\le L\le l_1+l_2$ applies. Due to the parity conservation we deduce $L+l_1+l_2$ is even. M is the (total) magnetic quantum number. The radial functions $\psi^{LMS}_{l_1l_2}(r_1,r_2)$ derive from the equations

$$\begin{aligned}[E_1+E_2-h_{l_1}-h_{l_2}]\;&\psi^{LMS}_{l_1l_2}(r_1,r_2)\\ &-\sum_{l_1'l_2'}\left\langle l_1l_2\parallel\frac{1}{r_{12}}\parallel l_1'l_2'\right\rangle_L\psi^{LMS}_{l_1'l_2'}(r_1,r_2)=\chi^{LS}_{l_1l_2}(r_1,r_2)\end{aligned} \tag{3.2}$$

The single particle operators h_{l_j}, $j=1,2$ contain the electron–ion electrostatic potential and the centrifugal term, i.e.

$$h_{l_j}=-\frac{1}{2}\frac{\partial^2}{\partial r_j^2}-\frac{1}{r_j}+\frac{l_j(l_j+1)}{2r_j^2} \tag{3.3}$$

The function $\chi^{LS}_{l_1l_2}(r_1,r_2)$ has the form

$$\begin{aligned}\chi^{LS}_{l_1l_2}(r_1,r_2)&=\frac{i^L}{k_0}\sqrt{2\pi(2L+1)}\\ &\times\left[\left(\left\langle l_1l_2\left\|\frac{1}{r_{12}}\right\|0L\right\rangle_L-\frac{1}{r_2}\delta_{l_10}\delta_{l_2L}\right)\varphi(r_1)j(k_0r_2)+(-)^S(1\leftrightarrows 2)\right]\end{aligned} \tag{3.4}$$

where $j(x)$ is the Riccati–Bessel function and $\langle\cdots\left\|\frac{1}{r_{12}}\right\|\cdots\rangle$ is the reduced matrix element of the electron–electron Coulomb potential. The name exterior complex scaling refers to the fact that for the numerical solution of Eq. (3.2) on a finite grid one rotates at the distance R_0 the radial coordinates into the complex plane through an angle ϑ. This is done via the transformation

$$r_j'=r_j\text{ for }r_j<R_0\quad\text{and}\quad r_j'=R_0+(r_j-R_0)\mathrm{e}^{\mathrm{i}\vartheta}\text{ for }r_j\ge R_0,\quad j=1,2$$

Under this transformation, outgoing waves possess the asymptotic behavior $\psi_{l_1 l_2}^{LMS}(r'_1, r'_2) = 0$ for $r'_j \gg R_0$. Incoming waves diverge however under the rotation transformation and therefore $\chi_{l_1 l_2}^{LS}(r'_1, r'_2)$ must be truncated at $r'_j > R_0$. Numerically, Eq. (3.2) is solved using for example a 7-point finite difference method [80–83].

3.2 The Convergent Close Coupling Method

An accuracy comparable to that of the ECS method is achieved by the convergent close coupling method (CCC) [85, 98]. In the CCC approach the wave function of the electron–target system is expressed as an antisymmetric expansion using N square-integrable (L^2) wave functions φ_n^N $(n = 1, \ldots, N)$ with eigenvalues ϵ_n^N. These functions are obtained by diagonalizing the target Hamiltonian in a Laguerre basis and include states with negative and positive energies. When the number of elements of the basis set N is increased, the functions with a negative energy tend to the true discrete target eigenfunctions. The positive-energy pseudostates constitute an (with N) increasingly dense discretization of the target continuum. To set up uniquely the close-coupling (CC) equations one has to specify the N functions and the total energy E_i (cf. Eq. (1.4)) of the electron–target scattering system. In the CCC method these equations are presented as coupled Lippmann–Schwinger equations.

The final state $\Psi_{\boldsymbol{k}_1 \boldsymbol{k}_2, S}(\boldsymbol{r}_1, \boldsymbol{r}_2)$ in the CC method consists asymptotically of a plane wave $|\boldsymbol{k}_1\rangle$ for one electron and an L^2 state for the other excited electron $|\varphi_2^N\rangle$. This means that in the CCC theory ionization processes are treated as an excitation to the positive-energy target pseudostates. Physically, one may argue that the electron represented by ($|\varphi_2^N\rangle$) shields completely the residual ion field experienced by the electron which is then described by $|\boldsymbol{k}_1\rangle$. This shielding effect does not depend on the electron energies or positions. So, from this point of view, the CCC approach seems suitable for elastic and inelastic nonionizing scattering processes, nevertheless the CCC proved equally capable of describing ionization processes [85, 98], as illustrated below by a few examples.

Having solved the CC equations one then determines the transition matrix elements (1.3) and from these the appropriate cross sections.

A comparison between the numerical results for the cross section of the CCC and ECS methods can be found in Ref. [85].

3.3 Analytical Models

Another rather analytical methodology concentrates on finding approximate, explicit solutions of the (nonrelativistic) three-body final state $\Psi_{\boldsymbol{k}_1,\boldsymbol{k}_2}(\boldsymbol{r}_1,\boldsymbol{r}_2)$ [13, 14, 99, 113–117]. The initial state can be taken then as in Eq. (1.3) and the cross section Eq. (1.3) has to be evaluated, in general, numerically. In the previous chapter we discussed the wave functions $\Psi^{2\mathrm{C}}$ and $\Psi^{3\mathrm{C}}$ (Eq. (1.40)) which have been used in the context of ion–atom collisions [13] and electron-impact ionization [14, 99] with remarkable success (in comparison to the measured shapes of the angular distributions). The $\Psi^{3\mathrm{C}}$ amounts to viewing the system as a collection of three individual two-body subsystems, each treated to infinite orders (in a perturbative sense). As a consequence, the wave function possesses the correct asymptotic behavior at large interparticle separations and/or higher energies[1], meaning that asymptotically higher order corrections (beyond two-body scattering events) decay faster than the Coulomb potential.

The success of the 3C method sparked a debate about the importance of boundary conditions [14, 99–106] and some attempts have been made to modify standard theories so as to incorporate the proper asymptotic behavior. For instance, an interelectronic Coulomb phase factor has been included in the distorted-wave Born approximation (3DWBA) to restore the boundary conditions at large interparticle separations [105]. Similarly, an asymptotically correct FBA has been derived and employed for the calculation of cross sections [100]. In this context we recall our mathematical finding of Chapter 1, expressed in Eq. (1.23) which states that in the high energy regime ($E_j > \epsilon_i$, $j = 1, 2$) the behavior of the wave function Ψ at the origin ($\boldsymbol{r}_j = 0$) is decisive. The normalization of the wave function depends however on the behavior of the wave function in the entire configuration space. In fact the normalization of $\Psi^{3\mathrm{C}}$ is obtained from its asymptotic flux[2]. This realization is important for the use of various approximation of $\Psi_{\boldsymbol{k}_1,\boldsymbol{k}_2}(\boldsymbol{r}_1,\boldsymbol{r}_2)$ to calculate cross sections. For example, in Ref. [109] a wave function $\Psi^{(AM)}$ has been derived that is (to a leading order) correct in the entire asymptotic regime, i.e., for larger interparticle separations and when two particles approach each other whilst far away from the third one. This progress comes at the expense of introducing local relative momenta. The difficulty when attempting to calculate cross sections

[1] It is important to mention here that, as shown in [3], the spatial range of the asymptotic region grows linearly with increasing k_j, this is because the quantity that has to be large to attain the asymptotic limit is $k_j r_j$.

[2] More precisely, one requires the flux density generated by the asymptotic form of $\Psi_{\boldsymbol{k}_1,\boldsymbol{k}_2}(\boldsymbol{r}_1,\boldsymbol{r}_2)$ to be the same as that generated by plane waves [107].

with $\Psi^{(AM)}_{\boldsymbol{k}_1,\boldsymbol{k}_2}(\boldsymbol{r}_1,\boldsymbol{r}_2)$ arises from the unknown normalization which, as stated above, is decisive in determining the cross sections. In Ref. [109] an expression for the normalization has been proposed, however with no justification. Hence, it is not clear whether an artifact in the normalization or in the wave function itself is the reason for the deviation from experiments when $\Psi^{(AM)}_{\boldsymbol{k}_1,\boldsymbol{k}_2}(\boldsymbol{r}_1,\boldsymbol{r}_2)$ is employed for the evaluation of the cross sections [106].

3.3.1 Dynamical Screening

The approximate wave function $\Psi^{3C}_{\boldsymbol{k}_1,\boldsymbol{k}_2}(\boldsymbol{r}_1,\boldsymbol{r}_2)$ accounts correctly for two-body scattering events but does not include three-body and higher-order interactions, as discussed in full details in [3]. A strategy to partially remedy this shortcoming is to encompass these higher-order interactions as a dynamical screening of the two-body scattering properties. For example, in the dynamically screened 3C (DS3C) model the wave function $\Psi^{\mathrm{DS3C}}_{\boldsymbol{k}_1,\boldsymbol{k}_2}(\boldsymbol{r}_1,\boldsymbol{r}_2)$ has formally the same mathematical structure as in Eq. (1.40), however the charges that enter the Sommerfeld parameters are rather complicated functions of the positions. A full discussion of how these charges are determined and how the position dependence of the charges is converted into a momentum dependence is discussed in [3]. Here we just cite the functional form of these charges in their momentum dependent form: The Sommerfeld parameters in the DS3C model for two electrons moving in the field of a residual positive ion of unit charge are given by

$$\alpha_j^{\mathrm{DS3C}} = \frac{Z_j^{\mathrm{DS3C}}}{v_j}, \quad j \in \{1,2,12\} \tag{3.5}$$

where v_j are the velocities corresponding to the momenta $\boldsymbol{k}_j$ and Z_j^{DS3C}, $j \in \{1,2,12\}$ are the electron–nucleus and the electron–electron product charges, respectively. The velocity-dependent product charges are given by [110]

$$Z_{12}^{\mathrm{DS3C}}(\boldsymbol{v}_1,\boldsymbol{v}_2) = \left[1-(F\,G)^2\,a^{b_1}\right]a^{b_2} \tag{3.6}$$

$$Z_{1}^{\mathrm{DS3C}}(\boldsymbol{v}_1,\boldsymbol{v}_2) = -1+(1-Z_{12}^{\mathrm{DS3C}})\frac{v_1^{1+a}}{(v_1^a+v_2^a)v_{12}} \tag{3.7}$$

$$Z_{2}^{\mathrm{DS3C}}(\boldsymbol{v}_1,\boldsymbol{v}_2) = -1+(1-Z_{12}^{\mathrm{DS3C}})\frac{v_2^{1+a}}{(v_1^a+v_2^a)v_{12}} \tag{3.8}$$

The functions entering in Eqs (3.6) and (3.7) are defined as

$$F = \frac{3+\cos^2 4\alpha}{4}, \qquad \tan\alpha = \frac{v_1}{v_2} \tag{3.9}$$

$$G = \frac{v_{12}}{v_1+v_2} \tag{3.10}$$

$$b_1 = \frac{2v_1v_2\cos\Theta_k/2}{v_1^2+v_2^2} \tag{3.11}$$

$$b_2 = G^2(-0.5+\mu) \tag{3.12}$$

$$a = \frac{E}{E+05} \tag{3.13}$$

The energy E is measured in atomic units and $\mu = 1.127$ is the so-called Wannier index [21]. The interelectronic relative angle Θ_k is given by $\Theta_k = \cos^{-1}\hat{\boldsymbol{v}}_1\cdot\hat{\boldsymbol{v}}_2$.

The physical meaning of the dynamical charges is readily inferred from their functional forms (3.6)–(3.8). If two particles have comparable velocities they experience their full two-body Coulomb interactions. The third particle feels a net charge equal to the sum of the charges of the two "composite" particles. The wave function $\Psi^{\mathrm{DS3C}}_{\boldsymbol{k}_1,\boldsymbol{k}_2}(\boldsymbol{r}_1,\boldsymbol{r}_2)$ and its normalization then have just the same form as, respectively, Eq. (1.40) and Eq. (1.41) with the Sommerfeld parameters being replaced by those given in Eqs. (3.5).

The cross sections evaluated with the wave function $\Psi^{\mathrm{DS3C}}_{\boldsymbol{k}_1,\boldsymbol{k}_2}(\boldsymbol{r}_1,\boldsymbol{r}_2)$ turned out to be in satisfactorily good quantitative agreement with the experimental absolute and relative angular distributions within a wide range of energies [76, 85, 111] (some examples are given below). The aim here is however not to discuss the strengths or shortcomings of the methods mentioned above but to highlight the physics of two-electron emission. Only the results for some of the methods discussed above will be utilized for the purpose of analysis.

Comparisons of the results of the CCC , 3C, 3DWBA and PSCC methods for the ionization cross sections are given in Ref. [98] while in Ref. [106] the 3C cross sections are contrasted with those of the $\Psi^{(\mathrm{AM})}$ approximation and the DS3C approaches. A comparison between the ECS and the CCC is found in Ref. [85].

3.3.2 Influence of the Density of Final States

According to Eq. (1.26), a computationally simple way to obtain an insight into the behavior of the cross section consists in augmenting the cross sections obtained from the orthogonalized plane-wave approximation OPWA with $\overline{DOS}$, the correct final state (two-particle) density per

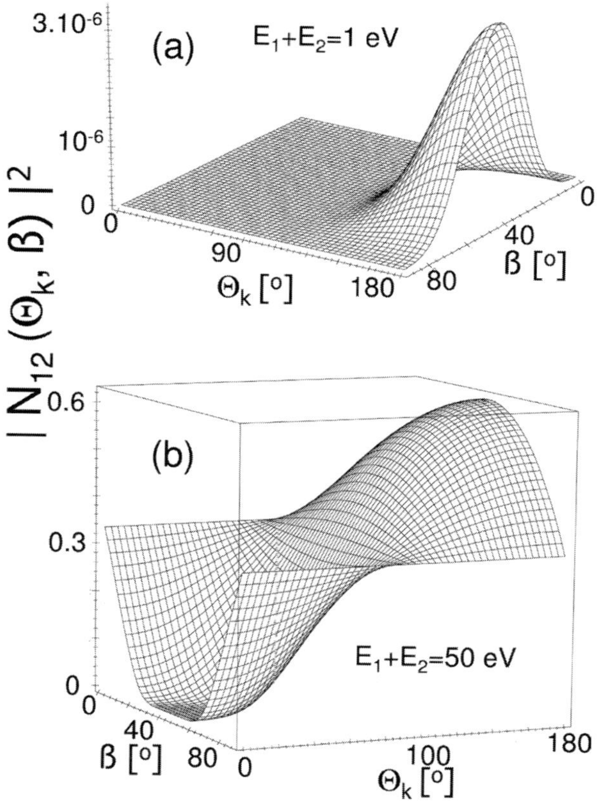

Figure 3.1: The electron–electron Coulomb density-of-states per unit volume, unit energy and unit momentum $|N_{12}(\Theta_k, \beta)|^2$, as given by Eq. (1.41). The angles Θ_k and β are given by Eq. (3.15). In (a) the total energy of the two electrons is $E_1 + E_2 = 1$ eV whereas in (b) $E_1 + E_2 = 50$ eV.

unit energy, unit volume and per unit momentum. This procedure becomes more accurate for higher energies, as follows from the mathematical analysis leading to Eq. (1.26). Nevertheless, at lower energies Eq. (1.26) provides an efficient tool to analyze analytically the qualitative trends in the cross sections. Hence we investigate here briefly some aspects of $\overline{DOS}(\boldsymbol{k}_1, \boldsymbol{k}_2)$.

In principle $\overline{DOS}$ depends on the details of the three-body final state and hence there exist no explicit expression for this quantity. For the 3C and DS3C wave functions $\overline{DOS}$ reads (note however that N_j, $j = 1, 2, 12$ are different for 3C and DS3C)

$$\overline{DOS} \quad = \quad |N_1 N_2 N_{12}|^2 \tag{3.14}$$

For higher energies ($Z/k_1 \to 0$, $Z/k_2 \to 0$, and $1/k_{12} \to 0$) we conclude that $\overline{DOS}$ tends to unity. Generally, the high-energy behavior of $\overline{DOS}$ is expected to be accurate because the corresponding wave functions (3C and DS3C) are high-energy approximations.

At low energies the generic behavior of $\overline{DOS}$ is most transparent when $\boldsymbol{k}_1, \boldsymbol{k}_2$ are parametrized as follows (the azimuthal angles of the momenta remain unchanged and in addition we assume the residual final ion to be unpolarized):

$$K = (k_1^2 + k_2^2)/2, \ \tan\beta = \frac{k_1}{k_2}, \ \text{and} \ \cos\Theta_k = \hat{\mathbf{k}}_1 \cdot \hat{\mathbf{k}}_2 \tag{3.15}$$

In this parametrization and for low energies $\overline{DOS}$ associated with the 3C wave function (i.e., Eq. (3.14)) reads for small K

$$|N_j|^2 = 2\pi \frac{Z}{k_j}; \quad j = 1, 2 \tag{3.16}$$

$$|N_1|^2 = \frac{1}{\sqrt{E}} \frac{2\pi Z}{\sqrt{2}\cos\beta}, \qquad |N_2|^2 = \frac{1}{\sqrt{E}} \frac{2\pi Z}{\sqrt{2}\sin\beta} \tag{3.17}$$

$$|N_{12}|^2 = \frac{2\pi}{\sqrt{2E} f_{\text{ee}}(\Theta_k, \beta)} \exp\left(-2\pi \frac{1}{\sqrt{2E} f_{\text{ee}}(\Theta_k, \beta)}\right)$$

$$f_{\text{ee}}(\Theta_k, \beta) = \sqrt{1 - \sin 2\beta \cos\Theta_k} \tag{3.18}$$

From these equations we infer that the angular behavior of $\overline{DOS}(\boldsymbol{k}_1, \boldsymbol{k}_2)$ is determined by[3] $|N_{12}|^2$ and hence the cross section vanishes exponentially when $\boldsymbol{k}_1$ approaches $\boldsymbol{k}_2$. At low energies the regime where $|N_{12}|^2$ practically vanishes extends substantially. Figures 3.1(a) and (b) illustrate how the angular dependence of $\overline{DOS}(\boldsymbol{k}_1, \boldsymbol{k}_2)$ changes with energy. In (a) ($K = 1$ eV) $\overline{DOS}(\boldsymbol{k}_1, \boldsymbol{k}_2)$ is practically finite only if $\boldsymbol{k}_1 = -\boldsymbol{k}_2$, whereas in (b) ($K = 50$ eV) we observe a strong dependence on β. For asymmetric energies ($\beta \approx 0°$, $90°$) $|N_{12}(\Theta_k)|^2$ depends weakly on Θ_k. In contrast, for $\beta \approx 45°$, i.e., for comparable energies $|N_{12}(\Theta_k)|^2$ is sharply peaked at $\hat{\boldsymbol{k}}_1 = -\hat{\boldsymbol{k}}_2$ and diminishes when $\hat{\boldsymbol{k}}_1 = \hat{\boldsymbol{k}}_2$ is approached.

[3] The cross section (1.2) as a function of the spherical angles $\Omega_{\boldsymbol{k}_1}, \Omega_{\boldsymbol{k}_2}$ and energies E_1 and E_2 associated with $\boldsymbol{k}_1$ and $\boldsymbol{k}_2$ is written as

$$\mathrm{d}\sigma(\Omega_{\boldsymbol{k}_1}, \Omega_{\boldsymbol{k}_2}, E_1, E_2) = (2\pi)^4 \frac{k_1 k_2}{v_0} |T|^2 \,\delta(E_{\text{f}} - E_{\text{i}}) d^2\Omega_{\boldsymbol{k}_1} d^2\Omega_{\boldsymbol{k}_2} \mathrm{d}E_1 \mathrm{d}E_2 \tag{3.19}$$

Hence, at low energies $|N_{12}|^2$ dictates the behavior of the cross section. For example the relation (1.26) for the cross section reads

$$\mathrm{d}\sigma(\Omega_{\boldsymbol{k}_1}, \Omega_{\boldsymbol{k}_2}, E_1, E_2) \approx (2\pi)^{12} \frac{Z^2}{v_0} |N_{12}|^2 \,|T_{\text{pw}}(\boldsymbol{k}_1, \boldsymbol{k}_2|^2 \,d^2\Omega_{\boldsymbol{k}_1} d^2\Omega_{\boldsymbol{k}_2} \mathrm{d}E_1 \mathrm{d}E_2 \tag{3.20}$$

3.4 Analysis of the Measured Angular Distributions

The nature of the physical mechanisms that underly the structures observed in the cross sections depends strongly on the energies and the emission angles of the two electrons. Here we discuss two prominent examples: The intermediate and the low energy regimes in the coplanar geometry as depicted in Fig. 1.1. The high-energy regime we considered in Figs. 1.1 and 1.3. The transition between these two regimes is gradual and is marked by the appearance of features in the cross sections which are akin to low-energy two-electron escape.

3.4.1 The Intermediate Energy Regime

Figure 3.2 shows the energy and the angular-resolved cross section for the electron-impact ionization of atomic hydrogen, evaluated according to Eqs. (2.29) and (1.2). This cross section is conventionally called triply differential cross section, TDCS, because it depends on the polar emission angles θ_1 and θ_2 of the electrons and on the energy E_2 of one of the electrons (the energy of the second electron is determined according to Eq. (1.4)). The azimuthal emission angles are fixed such that the two electrons are emitted in one plane (a schematics is shown in Fig. 1.1). The incident energy is $E_0 = 150$ eV which is much lower than in the case of Fig. 1.1. Nevertheless, in Fig. 3.2 we see the same structures in the measured and calculated cross sections as observed and analyzed in the situation of Figs. 1.1 and 1.2. In particular, the variations in the cross sections when changing the scattering geometry from Fig. 3.2(a) through (d) are readily understood by noting that the momentum transfer q becomes gradually smaller from Fig. 3.2(a) through (d). Hence, in the case of the smallest q the optical limit is approached and we observe the loss of sensitivity of the scattering process to the incoming beam direction $\hat{\boldsymbol{k}}_0$, i.e., the cross section shows, as in linear-polarized photoionization (with respect to the photon-beam direction), a backward–forward symmetry with respect to $\boldsymbol{k}_0$. With increasing q this symmetry is broken and the remaining dominant structure seen in Fig. 3.2a is the binary peak which we discussed in full detail in the preceding chapter.

Thus, we conclude that the cross section in this scattering geometry is determined by three quantities: The form factor of the potential which imposes the occurrence of a binary peak and pins down its overall position; the symmetry of the initial state which determines the shape of the binary peak; and the symmetry of the transition dipole operator (which implies that in the

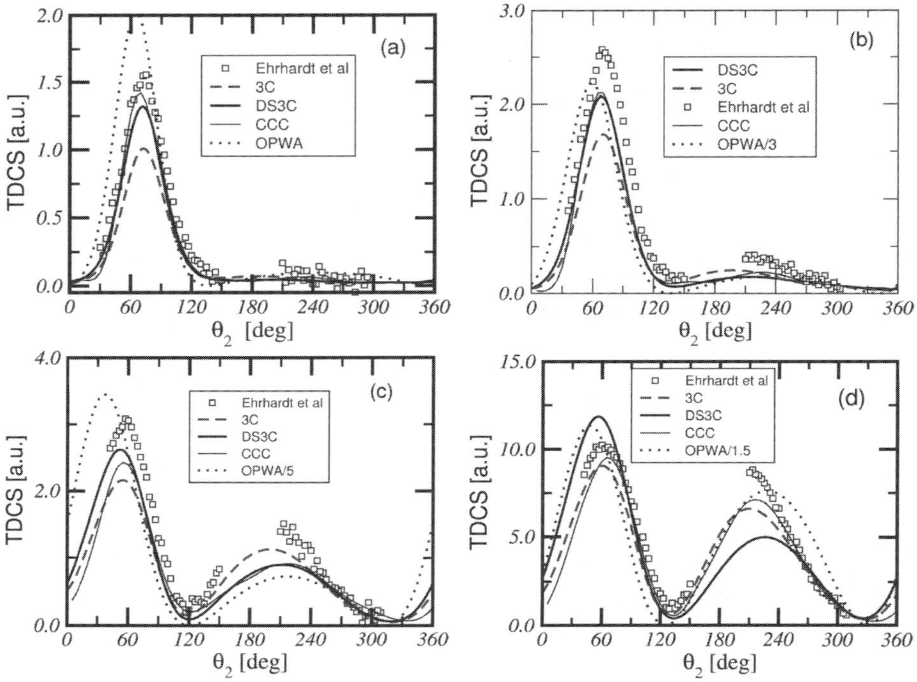

Figure 3.2: The energy and the angular-resolved cross sections (triply differential cross section, TDCS) for the coplanar electron-impact ionization of atomic hydrogen. In (a) one (fast) electron is detected under a fixed angle of $\theta_1 = -16^\circ$ with respect to the incident direction $\boldsymbol{k}_0$ (cf. Fig. 1.1). The cross section is then scanned as a function of the angle θ_2 of the other (slow) electron which has a fixed energy of $E_2 = 10$ eV. The incident energy is $E_0 = 150$ eV. Experimental data are taken from Ref. [75]. The solid thick curve stands for the DS3C results. The CCC calculations [98] yield the light solid curve. The 3C results are shown by the dashed curve. The dotted curves are the results for the cross sections when the orthogonalized plane wave approximation (OPWA), i.e., Eq. (1.16) is used. In (b) the angle of the fast electron is fixed to $\theta_1 = -10^\circ$. In (c) $\theta_1 = -4^\circ$, and in (d) $E_2 = 3$ eV and $\theta_1 = -4^\circ$. The OPWA results are scaled down by the factors depicted in the legend.

optical limit $\hat{\boldsymbol{k}}_0$ enters bilinearly in the TDCS). These general aspects are included in all the models we discussed so far and hence all the theories shown in Fig. 3.2(a–d) reproduce well the measured cross sections. In fact, the shape of the cross sections is well described even by the orthogonalized plane wave approximation (OPWA). On the other hand, OPWA leads to substantial deviations in the absolute magnitudes of the measured cross sections.

To understand why the magnitudes of the cross sections are more sensitive to the used theoretical models than the shapes we recall that, according to Eq. (1.26), the OPWA cross

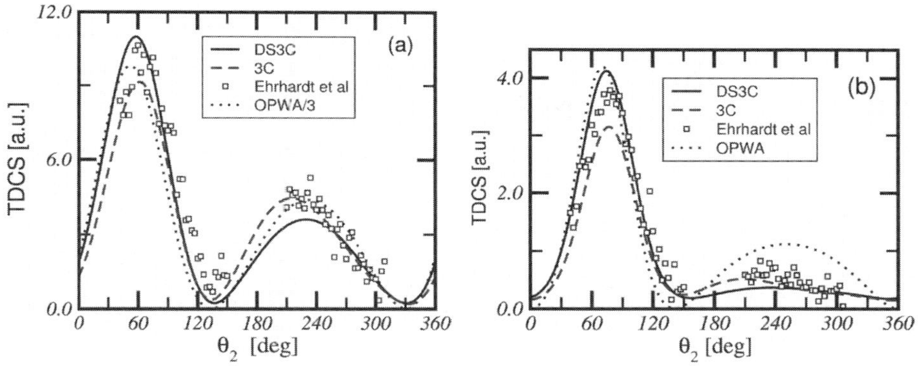

Figure 3.3: The same kinematical arrangements, i.e., coplanar geometry, as in Fig. 3.2, but the incident energy is increased to $E_0 = 250$ eV. In (a) $E_2 = 5$ eV and $\theta_1 = -8°$ and in (b) $E_2 = 5$ eV and $\theta_1 = -3°$. The curves represent the results of the same models as in Fig. 3.2.

sections have to be multiplied by $\overline{DOS}$. $\overline{DOS}$ is however not restricted by the symmetry of the transition matrix elements (which determines the shape of the cross section) but depends on the details of the three-body final state. In the case of Fig. 3.2 the OPWA have to be multiplied by $\overline{DOS}$, as given in Eq. (3.14) (which shows only a weak angular dependence in the situation of Figs. 3.2). Recalling the properties of $\overline{DOS}$ which we discussed in Fig. 3.1, it is then clear that the effect of $\overline{DOS}$ is a suppression of the OPWA cross section and a slight deformation of the angular distributions so as to decrease the cross section around the angular positions of the fixed electron.

This general behavior of the OPWA and the level of agreement with experiment and other more sophisticated theories persist when increasing the impact energy, as done in Fig. 3.3(a) and (b) ($E_0 = 250$ eV). In this figure we observe that the shapes of the cross sections are well reproduced by the OPWA, but not the absolute magnitudes (note also in this figure how the double-peak structure emerges when q becomes smaller).

3.5 Characteristics of the Correlated Pair Emission at Low Energies

As shown in Fig. 3.1 $\overline{DOS}(\boldsymbol{k}_1, \boldsymbol{k}_2)$ depends on the momentum vectors of the final state electrons and so plays a key role in determining the shape of the angular correlation within the electron pair, particularly at low energies (in the high energy limit $\overline{DOS}$ is constant). This is nicely demonstrated in Fig. 3.4 where the impact energy is lowered to $E_0 = 54.4$ eV. In Fig. 3.4 the momentum transfer q decreases from (a) through (d). Hence, the double-peak shapes of the cross sections calculated with OPWA become more emphasized when approaching (d). The fact that the OPWA calculations reproduce qualitatively the (double-peak) shapes and trends of the measured cross sections hints at the physical origins of these peaks being the same as in Figs. 3.2 and 3.3 that we discussed above.

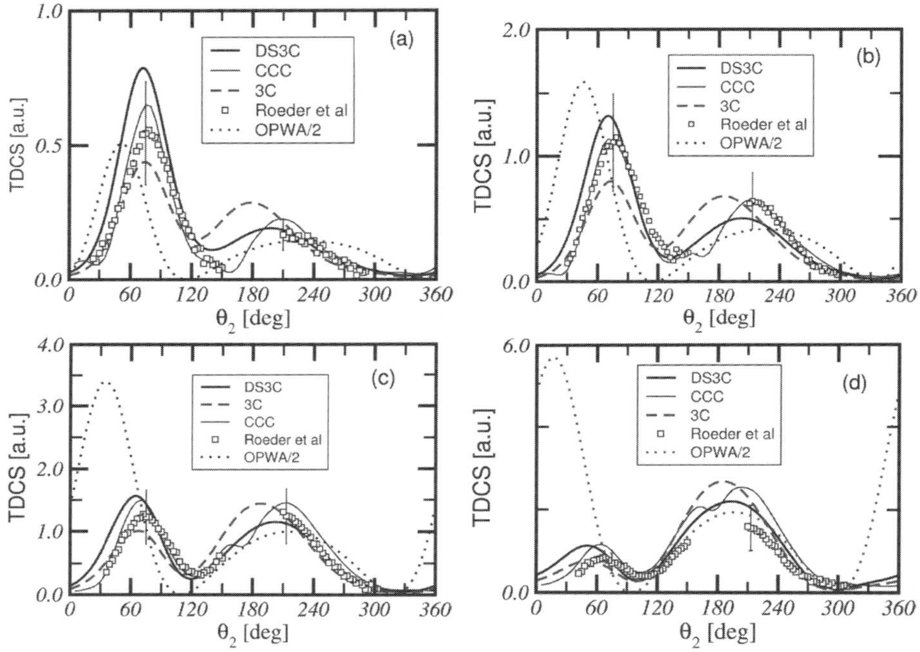

Figure 3.4: The TDCS for the ionization of atomic hydrogen in the same coplanar asymmetric geometry, as in Fig. 3.2. The incident energy is however decreased to $E_0 = 54.4$ eV. In (a) $E_2 = 5$ eV and $\theta_1 = -23^\circ$ whereas in (b) $E_2 = 5$ eV and $\theta_1 = 16^\circ$ and in (c) $E_2 = 5$ eV and $\theta_1 = -10^\circ$, and in (d) $E_2 = 5$ eV and $\theta_1 = -4^\circ$. Experimental data are due to Ref. [99]. The error bars indicate the uncertainty in the absolute value of the TDCS [85]. Curves are as in Fig. 3.2.

The obvious deviations of the OPWA from the precise measured peak values and positions and from other calculations are comprehensible upon considering the behavior of $\overline{DOS}$ associated with the 3C (or DS3C) wave functions (i.e., Eq. (3.14)): According to Eq. (1.26), the OPWA results have to be multiplied by $\overline{DOS}(\boldsymbol{k}_1, \boldsymbol{k}_2)$ (the resulting cross sections are not shown graphically for brevity). Recalling the interpretation of the behavior of $\overline{DOS}$ as shown in Fig. 3.1 we conclude that the overall magnitudes of the OPWA cross sections are decreased. Furthermore, the OPWA binary peaks in Fig. 3.4 are then shifted towards larger angles whereas the angular positions of the recoil peaks are pushed towards smaller angles. These changes are accompanied by a decrease in the cross section, particularly around the angular positions of the fixed electron, i.e., around $\theta_2 = \theta_1$. For example, in Fig. 3.4a $\theta_1 = -23°$ and hence we observe reduced cross sections around this angle, whereas in Fig. 3.4d $\theta_1 = -4°$ is almost in the forward direction which results in a substantial decrease and a shift in the (binary) peak in this case. All in all the situation presented in Fig. 3.4 marks the transition between the intermediate and the low energy regime in which a qualitatively different behavior of the cross section emerges.

3.5.1 Influence of the Exchange Interaction on the Angular Pair Correlation

An adequate analysis of the low-energy regime must include the influence of exchange, i.e., one has to address the question of which effects do appear in the transition amplitude (cf. Eq. (2.38)) due to the antisymmetry of the wave function[4]. Before studying the influence of the spin on the behavior of the spin-averaged quantities it is instructive to illustrate the general features of the exchange-induced effects which are quantified by the spin asymmetry, as introduced by Eq. (2.36) and (2.38). As mentioned and analyzed in the previous chapter, generally there are other sources for the spin dependence of the cross sections. However, in this work the discussion is restricted to exchange-induced effects.

As was pointed out in Chapter 2, the spin asymmetry A^s diminishes for $E_1 \gg E_2$ and at low energies it is anticipated by the PWA to be positive (cf. Fig. 2.3 and the associated discussion). In addition, if the initial target and the final ion state are not (orbitally) polarized, then we expect the triplet scattering to vanish, i.e.,

$$T^{(S=1)} = 0 \Rightarrow A^s = 0 \quad \text{if} \quad E_1 = E_2 \quad \text{and} \quad \hat{\boldsymbol{k}}_1 \cdot \hat{\boldsymbol{k}}_0 = \hat{\boldsymbol{k}}_2 \cdot \hat{\boldsymbol{k}}_0 \qquad (3.21)$$

These anticipations are confirmed experimentally by Fig. 3.5 which shows the dependence of A^s on the sharing of the excess energy $E = E_1 + E_2$ between the electrons. For a comparison the corresponding theoretical results within the 3C and the DS3C are also included. We recall that in this situation A^s does not depend on $\overline{DOS}$. Hence the deviations between the 3C

[4] It suffices to antisymmetrize either the initial or the final-state wave function.

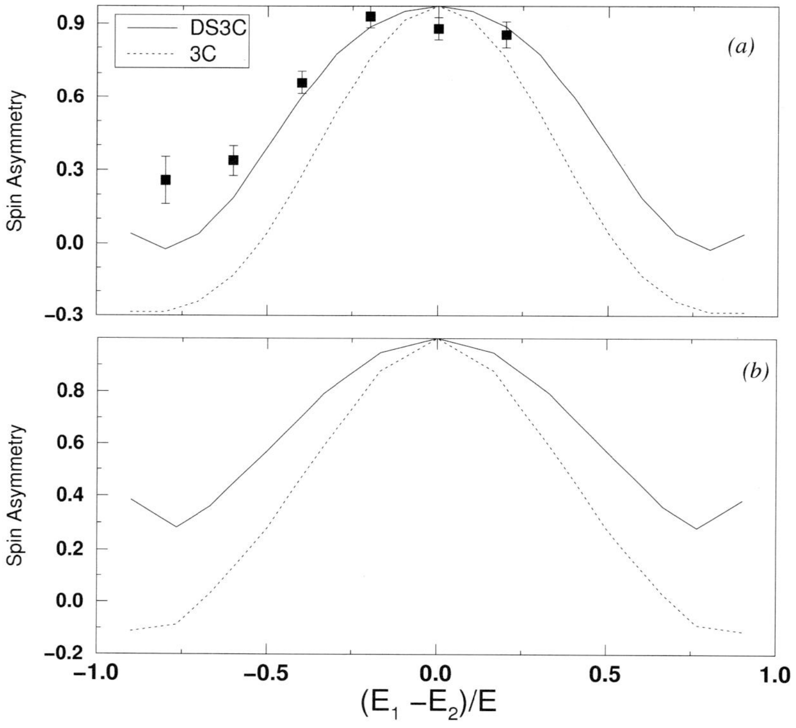

Figure 3.5: The spin asymmetry A^s in the (polarized) electron-impact ionization of the (polarized) valence electron of Li. The total kinetic energy of the emitted electrons is $E = E_1 + E_2 =$ 20 eV. Both of the electrons are detected co-planar with the incident direction $\boldsymbol{k}_0$. $\Theta_k = 90°$ whereby $\boldsymbol{k}_0$ bisects the inter-electronic relative angle Θ_k, i.e., in the notation of the top panel of Fig. 1.1 $\theta_1 = 45° = \theta_2$. The 3C (dotted curve) and DS3C (solid curve) results are shown along with the experimental findings (full squares) [112]. Theoretical results do not account for the finite experimental resolution. Inset (b) shows the same quantities as in (a) however the excess energy is lowered to $E = 6$ eV (no experimental data are available at this energy).

and the DS3C calculations stem from the different behavior of the radial part of the wave functions.

Inspecting Fig. 3.5(a) and (b) we conclude that the general behavior of A^s is rather insensitive to the excess energy and that the 3C wave function will not be capable of describing exchange-induced effects on the averaged cross section. These findings are also in line with our discussion of Fig. 2.3 where we find out that the 3C and the FBA yield inaccurate results for the momentum-integrated A^s in the low energy regime whereas the 2C (and the DS3C [110,113]) model performs well. Concerning Fig. 3.5(a) and (b) it should be mentioned that the use of the 2C wave function to evaluate A^s yields results that compare favorably with experiment (similar to the DS3C results) whereas the FBA results are rather comparable with the 3C predictions.

3.6 Threshold Behavior of the Energy and the Angular Pair Correlation

The low-energy behavior of the energy and the angular correlation within the electron pair is the result of an interplay of a variety of phenomena which we shall, to some extent, disentangle in this chapter. For this purpose the results of the PWA near threshold serve as a rough but useful guide. To keep the discussion general it is instructive to consider the problem without reference to a certain shape of the scattering potential.

3.6.1 Generalities of Threshold Pair Emission

The general expression of the T matrix elements in momentum space is given by Eq. (1.21) which indicates that the properties of the spectral density of the initial state are decisive, as they set the range of contributing momentum components of the final-state wave function and of the scattering potential. For a qualitative understanding of Eq. (1.21) at low energies let us assume that the final-state momentum-space wave function $\tilde{\Psi}^*_{\boldsymbol{k}_1\boldsymbol{k}_2}(\boldsymbol{p}_1, \boldsymbol{p}_2)$ is peaked[5] around $\boldsymbol{k}_j = \boldsymbol{p}_j, j = 1, 2$ (this is strictly valid at higher energies). More precisely, the assumption is that in Eq. (1.21) $\boldsymbol{p}_j$ vary within the interval $\boldsymbol{k}_j = \boldsymbol{p}_j \pm \Delta\boldsymbol{p}$, $j = 1, 2$, where for the constant vector $\Delta\boldsymbol{p}$ the relation $|\Delta\boldsymbol{p}| \ll |\boldsymbol{k}_j|$ applies. On the other hand, due to energy conservation, k_j cannot exceed $\sqrt{2(E_0 - \epsilon_i)}$ which tends to zero near threshold. Therefore, we can express the form factor in Eq. (1.21) in a Taylor series around $\boldsymbol{p}_1$, as done in Eq. (2.45), and cast

[5]To the knowledge of the author there is as yet no general (in the entire momentum space) proof of this statement in the case of many-body wave functions. For the two-body momentum-space wave function $\tilde{\psi}_{\boldsymbol{k}}(\boldsymbol{p})$ we infer from Eq. (1.39) (and the associated footnote) that $\tilde{\psi}_{\boldsymbol{k}}(\boldsymbol{p})$ is peaked around $\boldsymbol{k} = \boldsymbol{p}$.

Eq. (1.21) in the form[6]

$$T(\boldsymbol{k}_1,\boldsymbol{k}_2) = \tilde{U}(\boldsymbol{k}_0)\int \mathrm{d}^3\boldsymbol{p}_1\mathrm{d}^3\boldsymbol{p}_2\,\tilde{\Psi}^*_{\boldsymbol{k}_1\boldsymbol{k}_2}(\boldsymbol{p}_1,\boldsymbol{p}_2)\tilde{\varphi}^*(\boldsymbol{k}_0-(\boldsymbol{p}_1+\boldsymbol{p}_2))$$
$$-\int \mathrm{d}^3\boldsymbol{p}_1\mathrm{d}^3\boldsymbol{p}_2\,\tilde{\Psi}^*_{\boldsymbol{k}_1\boldsymbol{k}_2}(\boldsymbol{p}_1,\boldsymbol{p}_2)\left[(\boldsymbol{p}_1\cdot\boldsymbol{\nabla}_{\boldsymbol{x}})\tilde{U}(\boldsymbol{x})\Big|_{\boldsymbol{x}=\boldsymbol{k}_0}\right]\tilde{\varphi}^*(\boldsymbol{k}_0-(\boldsymbol{p}_1+\boldsymbol{p}_2))+\ldots. \tag{3.22}$$

The integrals in this relation are restricted to the intervals $\boldsymbol{p}_j \in [\boldsymbol{k}_j - \Delta\boldsymbol{p}, \boldsymbol{k}_j + \Delta\boldsymbol{p}]$.

Hence, in the low energy regime, the cross section is determined to a first order (i.e., the first term in Eq. (3.22)) by a convolution $\langle\tilde{\Psi}_{\boldsymbol{k}_1\boldsymbol{k}_2}(\boldsymbol{p}_1,\boldsymbol{p}_2)|\tilde{\varphi}^*(\boldsymbol{k}_0-(\boldsymbol{p}_1+\boldsymbol{p}_2))\rangle_{\boldsymbol{p}_1,\boldsymbol{p}_2}$ of the initial state with the three-body continuum. On the other hand, for $k_0 \gg k_j,\ j=1,2$ the momentum-space wave function of the initial state $\tilde{\varphi}(\boldsymbol{k}_0-\boldsymbol{k}_1-\boldsymbol{k}_2)$ can also be Taylor expanded around $\boldsymbol{p}_j \approx \boldsymbol{k}_j$ which also constitutes a useful tool for analysis.

A further interesting result that follows from the leading term in Eq. (3.22) is that the singlet $T^{(S=0)}$ and the triplet $T^{(S=1)}$ scattering amplitudes have a completely different behavior near threshold, namely

$$T^{(S=1)} = \left\langle\tilde{\Psi}_{\boldsymbol{k}_2\boldsymbol{k}_1}(\boldsymbol{p}_1,\boldsymbol{p}_2)|U^t_{\text{eff}}|\tilde{\varphi}^*(\boldsymbol{k}_0-(\boldsymbol{p}_1+\boldsymbol{p}_2))\right\rangle_{\boldsymbol{p}_1,\boldsymbol{p}_2} \tag{3.23}$$

$$T^{(S=0)} = 2\tilde{U}(\boldsymbol{k}_0)\langle\tilde{\Psi}_{\boldsymbol{k}_1\boldsymbol{k}_2}(\boldsymbol{p}_1,\boldsymbol{p}_2)|\tilde{\varphi}^*(\boldsymbol{k}_0-(\boldsymbol{p}_1+\boldsymbol{p}_2))\rangle_{\boldsymbol{p}_1,\boldsymbol{p}_2}$$
$$-\left\langle\tilde{\Psi}_{\boldsymbol{k}_2\boldsymbol{k}_1}(\boldsymbol{p}_1,\boldsymbol{p}_2)|U^s_{\text{eff}}|\tilde{\varphi}^*(\boldsymbol{k}_0-(\boldsymbol{p}_1+\boldsymbol{p}_2))\right\rangle_{\boldsymbol{p}_1,\boldsymbol{p}_2} \tag{3.24}$$

$$U^t_{\text{eff}} = (\boldsymbol{p}_2-\boldsymbol{p}_1)\cdot\boldsymbol{\nabla}_{\boldsymbol{x}}\tilde{U}(\boldsymbol{x})\Big|_{\boldsymbol{x}=\boldsymbol{k}_0} \tag{3.25}$$

$$U^s_{\text{eff}} = (\boldsymbol{p}_2+\boldsymbol{p}_1)\cdot\boldsymbol{\nabla}_{\boldsymbol{x}}\tilde{U}(\boldsymbol{x})\Big|_{\boldsymbol{x}=\boldsymbol{k}_0} \tag{3.26}$$

From these equations we see that if the gradient of the potential $\tilde{U}(\boldsymbol{p})$ at the momentum $\boldsymbol{k}_0$ is small, singlet scattering dominates over triplet and hence A^s tends to unity. Furthermore, if $\tilde{\Psi}_{\boldsymbol{k}_1\boldsymbol{k}_2}(\boldsymbol{p}_1,\boldsymbol{p}_2)$ is peaked around $\boldsymbol{p}_j = \boldsymbol{k}_j$ we also find that triplet scattering vanishes in contrast to the singlet scattering amplitude. As far as the angular and energy behavior are concerned it is useful to analyze them for the Coulomb potential case and the conclusions for this potential can be verified for Eqs. (3.23) and (3.24).

[6]Note that Eq. (3.22) is valid when exchange effects are neglected (Eq. (3.22) corresponds to the direct term f). To evaluate the exchange term g one has to calculate, in addition to the integral in Eq. (2.45) the function $T(\boldsymbol{k}_2,\boldsymbol{k}_1)$, which is obtained by interchanging in Eq. (2.45) $\boldsymbol{k}_1$ and $\boldsymbol{k}_2$. Building then the sum and the difference of $T(\boldsymbol{k}_1,\boldsymbol{k}_2)$ and $T(\boldsymbol{k}_2,\boldsymbol{k}_1)$ one obtains respectively the singlet $T^{(S=0)}$ and the triplet $T^{(S=1)}$ scattering amplitudes.

3.6.2 Threshold Pair Emission from a Coulomb Potential

For the Coulomb potential Eqs. (3.23) and (3.24) read

$$T^{(S=1)} = \frac{1}{\pi^2}\left[\frac{\boldsymbol{K}^- \cdot \boldsymbol{k}_0}{k_0^4}\right]\tilde{\varphi}^*(\boldsymbol{k}_0 - \boldsymbol{K}^+) \tag{3.27}$$

$$T^{(S=0)} = \frac{1}{\pi^2}\left[\frac{1}{k_0^2} + \frac{\boldsymbol{K}^+ \cdot \boldsymbol{k}_0}{k_0^4}\right]\tilde{\varphi}^*(\boldsymbol{k}_0 - \boldsymbol{K}^+) \tag{3.28}$$

Here we introduced the interelectronic vector $\boldsymbol{K}^- = (\boldsymbol{k}_1 - \boldsymbol{k}_2)$. From Eqs. (3.27) and (3.28) we deduce several conclusions which will be of importance for the interpretation of the experimental findings:

1. Near threshold singlet scattering is dominant and A^s approaches unity, for K^- tends to zero near threshold.

2. For a spatially isotropic initial state the low-energy singlet scattering cross section shows only a weak angular dependence, since the leading term is $\left|\frac{\tilde{\varphi}(\boldsymbol{k}_0 - \boldsymbol{K}^+)}{k_0^2}\right|^2$.

3. The singlet scattering cross section depends only on $\boldsymbol{K}^+$. For $\boldsymbol{K}^+ = 0$ the singlet cross section is proportional to the momentum (Compton) profile of the initial state $|\tilde{\varphi}(\boldsymbol{k}_0)|^2$.

4. For a constant $\boldsymbol{K}^-$ the angular dependence of the spin-averaged cross section is akin to the behavior of the momentum-space wave function of the initial state. For example, for an initial s state, the cross section is maximal when $|\boldsymbol{k}_0 - \boldsymbol{K}^+|$ is minimal and vice versa. This situation changes qualitatively for a state with different symmetry (e.g., p or d state), as readily inferred from Eqs. (1.8)–(1.10).

5. In a spin-averaged experiment the triplet and the singlet cross sections can be measured separately without the need for a spin resolution as follows: one fixes $\boldsymbol{K}^+$ and varies $\boldsymbol{K}^-$. In this case the singlet cross section acts as a constant background for the spin-averaged cross section. The magnitude of this background is determined when, due to symmetry the triplet scattering cross section vanishes. This is the case when $\boldsymbol{K}^- = 0$ or when $\boldsymbol{K}^- \perp \boldsymbol{k}_0$

6. For a fixed $\boldsymbol{K}^+$, the angular dependence of the spin-averaged cross section stems from the triplet scattering and behaves as $\cos^2\theta_{\boldsymbol{K}^-}$, where $\cos\theta_{\boldsymbol{K}^-} = \hat{\boldsymbol{K}}^- \cdot \hat{\boldsymbol{k}}_0$

Some of the above conclusions can also be made in the general case of Eqs. (3.23) and (3.24). As we will show now vestiges of these rules appear even at energies as high as two times the ionization potential. The reason for this is the smoothness of the Coulomb potential (cf. Eqs. (3.23) and (3.24)).

3.6.3 Regularities of the Measured Pair Correlation at Low Energies

There is an impressive range of available experimental data for the spin unresolved cross sections (1.2) for atomic systems. Here we will concentrate and analyze in depth some typical examples, plotted in Fig. 3.6. It appears from this figure that the two-lobe structure that we encountered in the high and intermediate energy regime survives even at low energies. The underlying physics of these observations is, however, qualitatively different. First we notice here that the cross sections also decrease moderately with increasing θ_1, the angular position of the fixed electrons. This increase is accompanied by an increase in the momentum transfer q, as shown explicitly by the inset in Fig. 3.7, even though the change in q is rather small. From Eqs. (3.27) and (3.28) it is clear that at low energies the relevant quantity is not q but the recoil momentum $\boldsymbol{k}_{\text{rec}} = \boldsymbol{k}_0 - \boldsymbol{K}^+$, whose behavior as a function of θ_2 is shown in Fig. 3.7. As we see from this figure there is a specific angular position $\theta_2 \approx 340°$ where $|\boldsymbol{k}_{\text{rec}}|$ is minimal and a maximum of $|\boldsymbol{k}_{\text{rec}}|$ is located around $\theta_2 \approx 160°$. Hence, from Eq. (3.28) we expect the dependence on θ_2 of the singlet cross section to be due to that of the initial state $\tilde{\varphi}$ on $\boldsymbol{k}_{\text{rec}}(\theta_2)$. Since the initial state is an s state, $|\tilde{\varphi}(\boldsymbol{k}_{\text{rec}})|$ is maximal when $\boldsymbol{k}_{\text{rec}}(\theta_2)$ is minimal and vice versa. Hence from the structure of Eq. (3.28) we expect a single peak in the singlet cross section around $\theta_2 \approx 340°$ and a deep depression around $\theta_2 \approx 160°$ which is confirmed by the FBA cross section shown in Fig. 3.8b. Since the FBA is generally not appropriate for describing the spin asymmetry A^s, i.e. it weights wrongly the singlet and the triplet cross sections (cf. 2.3), we see in the FBA spin-averaged cross sections only the single-lobe singlet structure (note that in Fig. 3.8 the FBA triplet cross section is ca. 50 times smaller than the singlet). This is an essential point in so far as a multiple peak structure is indeed predicted by Eq. (3.27) to occur in the triplet scattering. According to Eq. (3.27) the angular structure dictated by the initial state (and manifested in the singlet scattering) has to be superimposed on the angular function $k_0(k_1 \cos\theta_1 - k_2 \cos\theta_2)$ (in Fig. 3.8 we have $k_1 = k_2$). The factor $k_0(k_1 \cos\theta_1 - k_2 \cos\theta_2)$ leads to a minimum in the triplet scattering at $\theta_2 = 0$

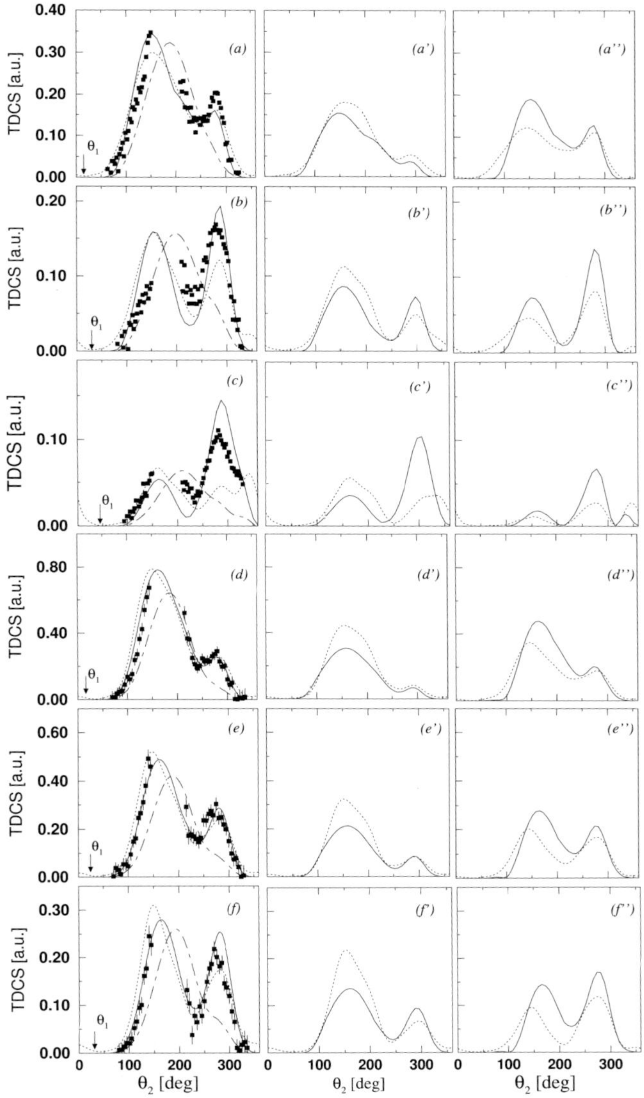

Figure 3.6: The spin-averaged TDCS for the ionization of atomic hydrogen in the same coplanar asymmetric geometry, as in Fig. 3.2. The incident energy is however decreased to $E_0 = 27.2$ eV. In $(a) - (c)$ $E_2 = 6.8 = E_1$ eV. The arrows indicate the positions of the fixed electrons, i.e., in (a) $\theta_1 = -15°$, in (b) $\theta_1 = -30°$ and (c) $\theta_1 = 45°$. The insets $(a') - (f')$ and $(a'') - (f'')$ show respectively the singlet and the triplet cross sections corresponding to the spin-averaged cross section shown in $(a) - (f)$. The DS3C (thick solid curve), 3C (dash-dotted curve), and CCC (dotted curve) are shown as well as the experimental data (full squares) [99]. The CCC results are multiplied by a factor of 1.99. The experiments are inter-normalized [118]. In the inset (d, e, f) E_2 is changed to $E_2 = 4$ eV. The fixed angles of the scattered electrons (d, e, f) are respectively $\theta_1 = 16°, 23°, 30°$. The inter-normalized experiments are the full squares [118]. The CCC results are multiplied by a factor of 0.88.

and a maximum at $\theta_2 = 180°$. When this behavior is combined with the behavior dictated by $|\tilde{\varphi}(\boldsymbol{k}_{\rm rec})|$ (a maximum at $\theta_2 \approx 340°$ and minimum at $\theta_2 \approx 160°$ a double-peak structure in the triplet cross section emerges (which is left–right asymmetric with respect to $\boldsymbol{k}_0$). In addition, we recall that the triplet scattering vanishes due to symmetry, as required by Eq. (3.21). This occurs at two angular positions in Fig. 3.8c which completely explains the angular behavior of the FBA cross section.

From the above analysis two general conclusions emerge:

1. The symmetry of the initial state govern the shape of the cross section.

2. At low energies it is decisive to describe correctly the spin asymmetry. For example the 3C model overemphasizes triplet scattering (cf. Fig. 2.3) in the case of Fig. 3.6 (the 3C results are not shown for brevity). On the other hand, as we explained above triplet scattering possesses a maximum around $\theta_2 = 180°$. This results also in a peak in the spin-averaged 3C cross sections around $\theta_2 = 180°$ which is at variance with experiment, i.e., from the above we can expect the spin asymmetry to be described wrongly by the 3C wave function.

3.6.4 Role of Final-state Interactions in Low-energy Correlated Pair Emission

Comparing Figs. 3.6 and 3.8 we notice a qualitatively different behavior between the singlet cross sections predicted by the FBA and other more accurate theories (DS3C and the CCC). To unravel the reason for this discrepancy and to point out a further important element of the physics of low-energy pair emission, we plotted in Fig. 3.9 the cross section calculated while treating the final state as prescribed by the 2C wave function, as given by Eq. (1.36). In general, when using methods more sophisticated than the FBA one has to account for the scattering of the projectile from the residual ion potential (1.20). The corresponding scattering amplitude vanishes due to the orthogonality of the target states in the case of FBA as well as when the 2C approximation is employed. However, when describing the final state continuum electrons by the 2C wave function one allows for scattering of both electrons from the residual ion potential via final-state interactions. It is this scattering which leads to the appearance of an additional peak (compared to the FBA) in Fig. 3.9 in the vicinity of the forward direction ($\theta_2 \approx 45°$). So, in Fig. 3.9, the peak around $\theta_2 \approx 45°$ is due to a corresponding maximum in the initial bound-state momentum distribution (so it depends on the symmetry of the initial state) and the peak around $\theta_2 \approx 45°$ is due to a distortion of the scattered electron wave by the residual ion and hence depends primarily on the charge of the ion Z and the energy E_1, but less so on the symmetry of the initial bound state.

Having pointed out the influence of final-state interactions it should be mentioned that the conclusions made in Section 3.6.2 apply to the 2C results: The shape of the angular correlation in the singlet channel is hardly changed (Fig. 3.10) when varying θ_1 whereas in the triplet channel the cross section depends strongly on the position of the fixed electrons (Fig. 3.11). The variations in the triplet cross section can be made comprehensible upon a similar analysis as in the case of the FBA results (Fig. 3.8c), i.e., they are the result of an interplay between the angular factor $\boldsymbol{k}_0 \cdot \boldsymbol{K}^-$, the propensity rules for triplet scattering, momentum distribution of the initial bound state, and final-state interactions between the scattered electron and the residual ion.

The missing element in the 2C approximation is the interaction between the final-state electrons. To incorporate this effect, at least on the level of correcting for the density of final states, we multiplied the 2C results by $|N_{12}|^2$. In this case the density of final states becomes that associated with the 3C model. As deduced from Figs. 3.8 and 3.6 the resulting cross sections agree remarkably well with experiment, in contrast to the 3C results, which hints at a shortcoming of the 3C wave function in the way the electron–electron scattering dynamics is treated as being in free space, i.e., without accounting for dynamical screening effects due to the presence of the residual ion. These defects are remedied partly by the DS3C model which indeed leads to results comparable in accuracy with full numerical calculations (cf. Fig. 3.6).

3.6.5 Interpretation of Near-threshold Experiments

As we have seen above, the conclusions of Section 3.6.2 were very helpful in identifying the physical origin of the structures seen in Fig. 3.6. The manifestation of these rules becomes even more evident in the vicinity of the threshold region. In particular, one can realize the experiment in such a way as to focus on the relevant dependences of the cross sections. For example, as evident from Eqs. (3.27) and (3.28) the structure in the cross sections is more comprehensible when plotted against $\boldsymbol{K}^-$ and/or $\boldsymbol{K}^+$. This is achieved for instance when the inter-electronic angle Θ_k is fixed. In this case K^- and K^+ are fixed. A prototypical,

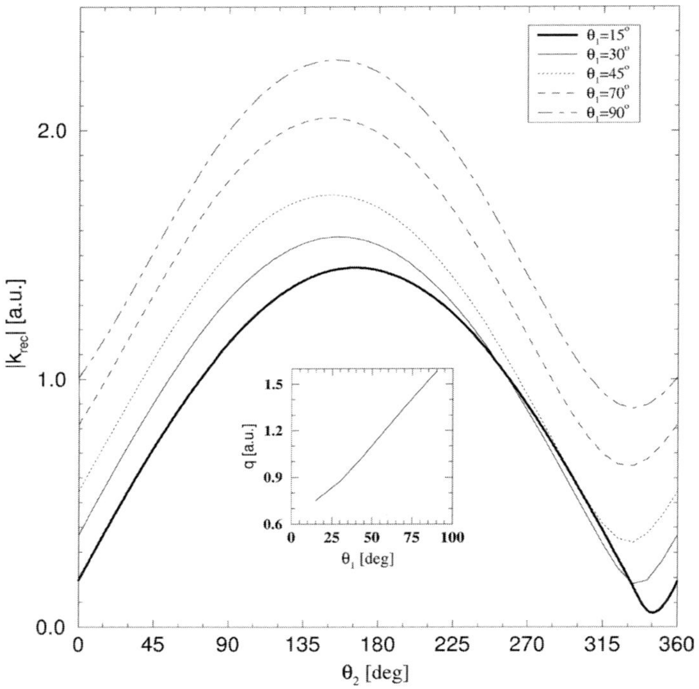

Figure 3.7: The recoil-ion momentum $k_{\rm rec}$, as determined from Eq. (1.5), plotted versus the emission angle θ_2 for the cases of Fig. 3.6(a)–(c). The inset shows the variation of the momentum transfer q with the angle θ_1.

spin-nonresolved case is shown in Fig. 3.12 where the magnitudes of the momenta $k_1 = k_2$ are fixed as well as the interelectronic angle $\Theta_k = 90°$. One then varies θ_2 as sketched in the inset of Fig. 3.12 ($\hat{\boldsymbol{K}}^-$ and $\hat{\boldsymbol{K}}^+$ are then both varied). At $\theta_2 = 45°$ (that is when $\boldsymbol{k}_2 \parallel \mathcal{A}$ in the inset) the vector $\boldsymbol{K}^+$ is parallel to $\boldsymbol{k}_0$ and hence the recoil momentum $\boldsymbol{k}_0 - \boldsymbol{K}^+$ is minimized. From Eqs. (3.27) and (3.28) we expect then a maximum in the cross section because the s state probability amplitude deceases with increasing momentum. However, due to the symmetry condition (3.21) the triplet scattering vanishes here. This does not imply necessarily that the spin non-resolved cross section should vanish at this point, however, as we stated in Section 3.6.2 the singlet scattering cross section exhibits only a weak angular dependence and hence, in the vicinity of threshold we observe the angular structures dictated by the threshold behavior of the triplet scattering cross section (3.27). This cross section possesses a maximum when $\boldsymbol{K}^-$ is parallel to $\boldsymbol{k}_0$, which occurs in Fig. 3.12 at the angular position denoted by $\mathcal{B}$ in

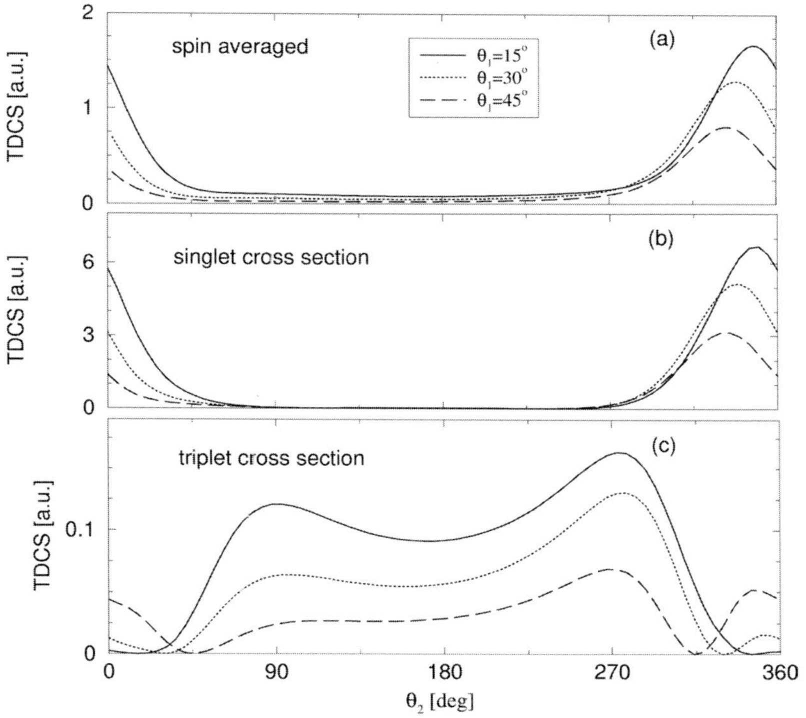

Figure 3.8: Panel (a) shows the spin-averaged cross section for the collision geometry of Fig. 3.6(a)–(c) as predicted by the FBA (Eq. 1.11). The corresponding singlet and triplet cross sections are depicted in (b) and (c), respectively.

the inset ($\theta_2 = 135°$). When $\boldsymbol{k}_2$ becomes parallel to $\mathcal{A}'$, the total momentum vector of the pair points in the backward direction and the recoil momentum is maximal. Equations (3.27) and (3.28) predict a minimum in the cross section if the initial momentum spectral density declines with increasing momentum (which is the case for s states). In addition, at this position, the triplet scattering cross section vanishes as well. When $\boldsymbol{k}_2$ is aligned along $\mathcal{B}'$ the inter-electronic vector $\boldsymbol{K}^-$ is antiparallel to $\boldsymbol{k}_0$ and the triplet cross section possesses a maximum.

These interpretations, which are based on the threshold formulas Eqs. (3.27) and (3.28) are nevertheless still viable even at considerably high energies. At $E_1 = E_2 = 25$ eV, only a very slight trace of the threshold behavior remains and the cross section now consists of a single lobe at $\theta_2 = 45°$, i.e., when the two electrons are positioned symmetrically and on different sides of $\boldsymbol{k}_0$. The emergence of this peak when the energy is increased, combined

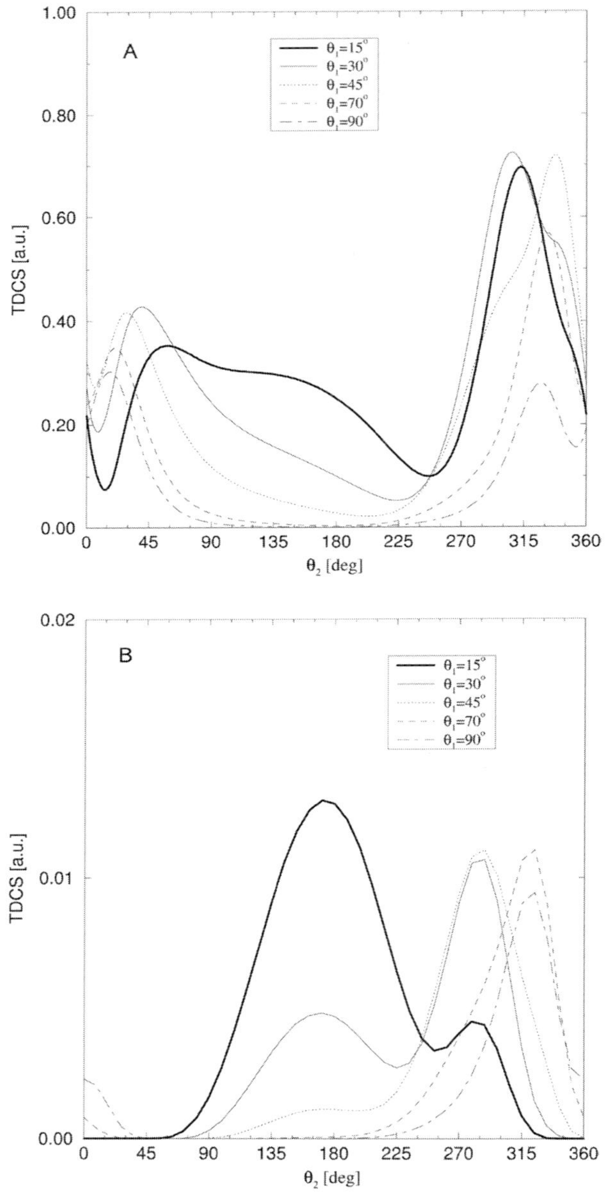

Figure 3.9: A: The same as in Fig. 3.8 (a), however more scattering angles θ_1 are included and the calculations are done using the 2C wave function for the final state, as given by Eq. (1.36). B: the same as in the upper panel, however, the cross sections of the upper panel are multiplied by the function $|N_{12}|^2$ that derives from Eq. (3.18).

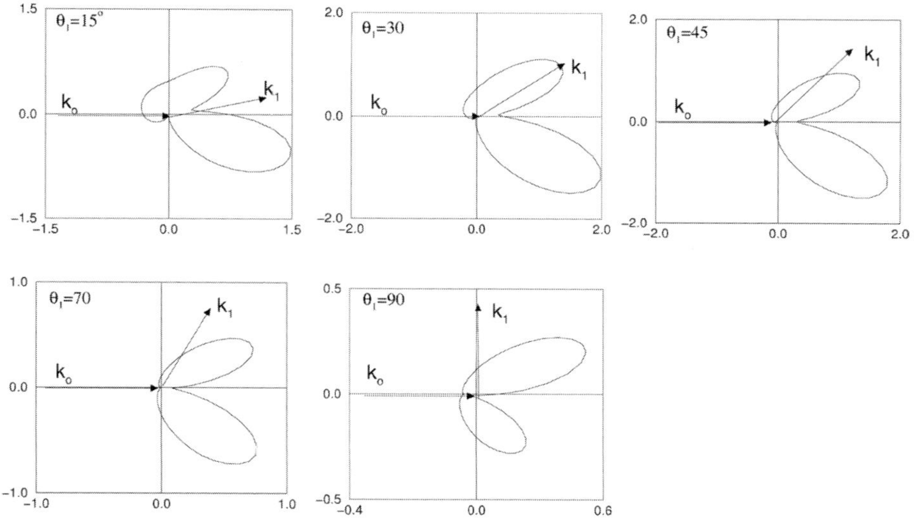

Figure 3.10: The polar plots show the singlet cross sections corresponding to Fig. 3.9. The angular position θ_1 of the fixed electron is denoted by the arrows and is also specified on the figures.

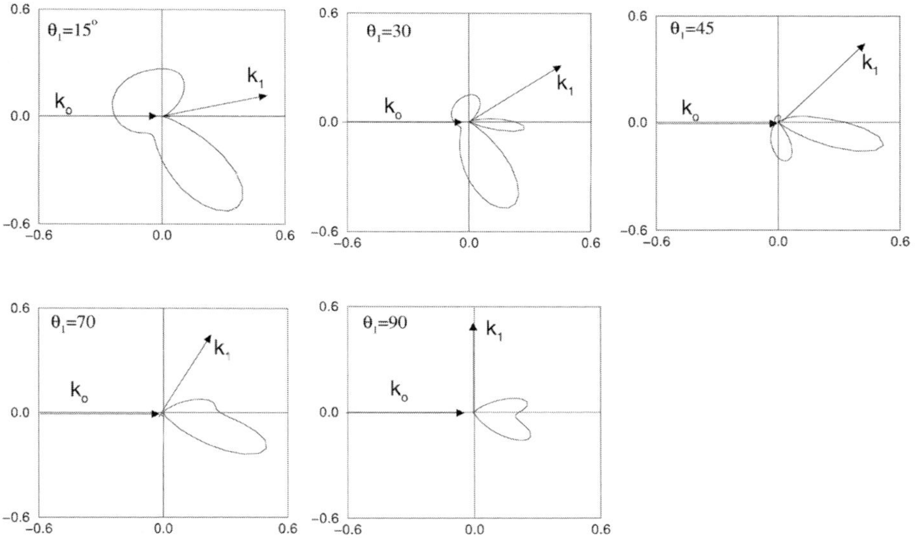

Figure 3.11: The same as in Fig. 3.9, however the triplet cross sections are plotted.

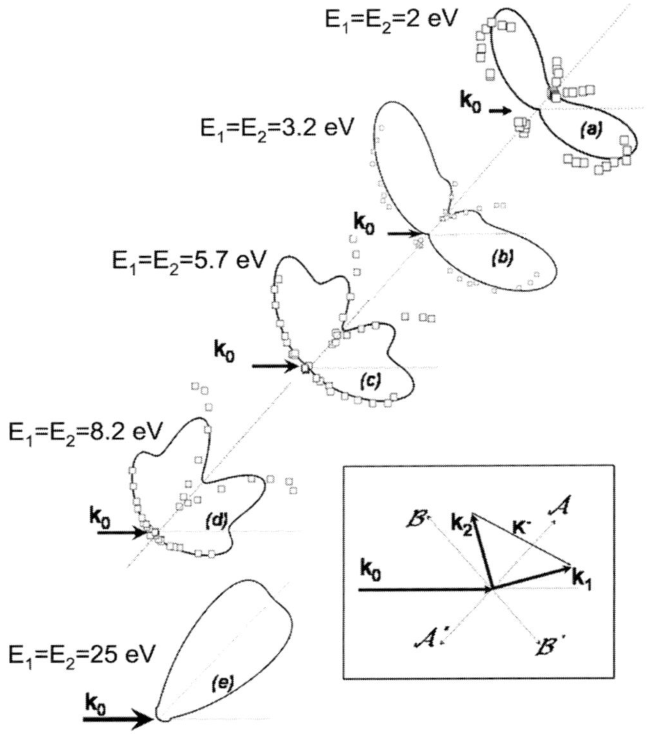

Figure 3.12: Polar plots for the triply differential cross section for the ionization of ground state atomic hydrogen in the coplanar, equal-energy, fixed-relative angle geometry. The two continuum electrons escape perpendicularly to each other ($\Theta_k = 90°$). As sketched in the inset, the plotted polar angle is θ_2, where $\cos\theta_2 = (\hat{\boldsymbol{k}}_2 \cdot \hat{\boldsymbol{k}}_0)$. In *(a)* through *(e)* the excess energies, $E_1 = E_2$ of the electrons are enhanced (by increasing E_0) by the amount indicated on the figures. Squares symbolize the relative experimental data. The TDCS is calculated using the DS 3C wave function for the final state. The theory predictions for the absolute values of the TDCS are as follows: in *(a)* TDCS$(\theta_2 = 135°) = 0.038$ a.u., in *(b)* TDCS$(\theta_2 = 135°) = 0.0525$ a.u., in *(c)* TDCS$(\theta_2 = 0°) = 0.1279$ a.u., in *(d)* TDCS$(\theta_2 = 45°) = 0.12$ a.u., and in *(e)* TDCS$(\theta_2 = 45°) = 0.01455$ a.u..

with the remanence of the low energy behavior, explains the occurrence of a shoulder-type structure in the cross section in the intermediate energy regime. The angular position for the peak in the case of $E_1 = 25$ eV is that for a direct (classical) electron–electron encounter. In our case it appears because it is favored by the form factor of the potential. So we see here the coincidence of two peaks with different origins. The low-energy singlet cross section possesses a peak at $\theta_2 = 45°$ because the initial-state momentum density is peaked at this

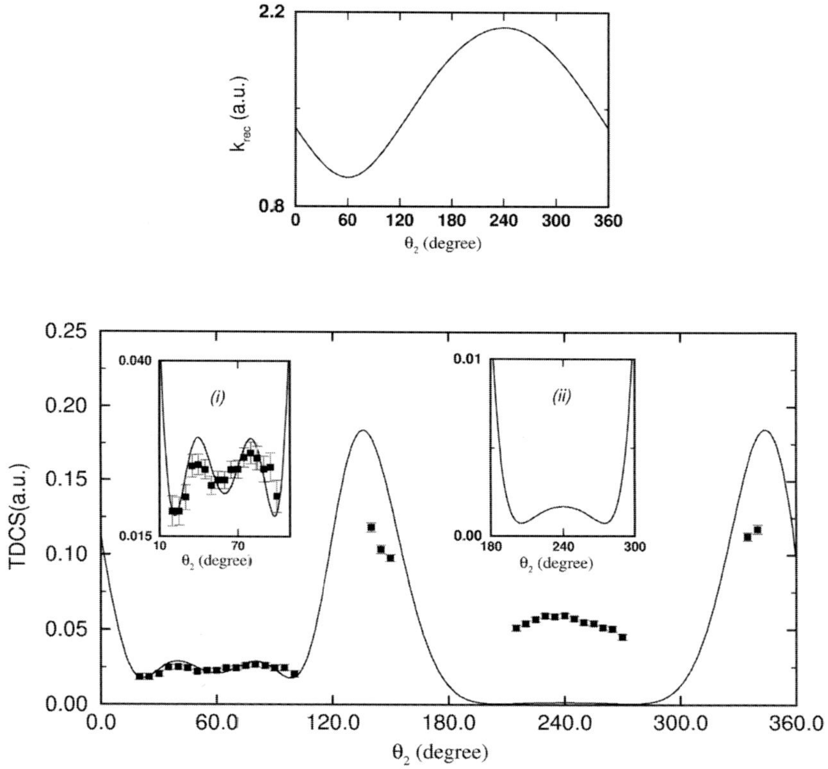

Figure 3.13: The TDCS for the electron-impact single ionization of the ground-state helium atom, He(^1S). As in Fig. 3.12 the inter-electronic angle is fixed, however in this case it is $\Theta_k = 120°$. The scattering geometry is otherwise as sketched in the inset of Fig. 3.12. The incident energy is fixed to $E_0 = 32.6$ eV (i.e., $E_1 = E_2 \approx 4$ eV). The absolute experimental data are represented by the squares. The insets show on an enlarged scale parts of the angular distribution whereas in the upper panel the angular behavior of the recoil-ion momentum k_{ion} is depicted.

angle, whereas at a high energy the origin of this peak lies in the form factor of the potential. The reason for this coincidence is the similar functional dependence in momentum space of the s state and the Coulomb potential. An initial state with a different symmetry is expected to lead to decisively different structures and trends than those observed in Fig. 3.12.

Another point worth exploring in this context is the independence (at very low energy) of the singlet cross section on the initial state when $\boldsymbol{K}^+ = 0$, as we pointed out in Section 3.6.2 (cf. Eq. (3.28)). Unfortunately there are as yet no experiments where the angular correlation has been studied for the same target but with different initial state while $K^+ = 0$. However, experiments have been performed where $\boldsymbol{k}_1 = -\boldsymbol{k}_2$ for the same target. In this case we expect, according to our discussion in Section 3.6.2, that the low-energy angular correlation will be dominated by triplet scattering. The singlet scattering serves as a background whose value is determined when triplet scattering is excluded. An example is shown in Fig. 3.13 for atomic helium as a target. Here K^+ is not exactly zero because $k_1 = k_2$, but $\Theta_K = 120°$. Nonetheless we expect the angular correlation pattern to be essentially that of the triplet scattering (Eq. (3.27)) which is basically confirmed by Fig. 3.13. At $\theta_2 = 60°$ the triplet cross section vanishes and at this point we see only the singlet channel electrons. Since $K^+ \neq 0$ the singlet scattering shows a flat peak at this point where k_{rec} is minimal (cf. upper panel of Fig. 3.13), which, if superimposed on the triplet cross section leads to the fine variation of the cross section in the regime around $\theta_2 = 60°$. At $\theta_2 = 150°$ and $\theta_2 = 330°$ the electron–electron relative momentum is parallel to $\boldsymbol{k}_0$ and we expect a maximum in the cross section, which is confirmed also by the full calculations. This suggests that the majority of the electron pairs associated with these two maxima belongs to the triplet channel.

At the angular position $\theta_2 = 240°$ triplet scattering vanishes as in the case where $\theta_2 = 60°$. The DS3C theory and the experiment show however a rather flat maximum (in an otherwise deep minimum). Contrasting the situation in this angular region with that around $\theta_2 = 60°$ we come to the conclusion that for the configuration around $\theta_2 = 240°$ the singlet scattering cross sections dominate, for a double-peak structure as around $\theta_2 = 60°$ would have been observed otherwise. The peak structure around $\theta_2 = 240°$ hints at a similar behavior in the singlet cross section which must be associated with processes not captured by the PWA.

3.7 Remarks on the Mechanisms of Electron-pair Emission from Atomic Systems

In addition to the mechanisms for the electron-pair emission we have discussed so far a number of other sources may influence strongly the angular correlations. Here we list some of them without detailed examples.

As we have seen in the case of isolated two-electron scattering, the energy and the momentum conservation laws impose certain final-state angular and energy patterns for two colliding particles. The structure of these (kinematical) pattern are easily deduced if the particles' initial momenta are known (otherwise we have to average over a distribution of initial momentum components which smears out the features predicted by the conservation laws). In particular, at high energies, one may assume the initial bound electron to be stationary on the scale of the passage time of the incoming projectile through the collision region. Hence the kinematical structures in the cross sections are more pronounced in the high-energy regime.

A few-body system can, to a certain extent, be viewed as a collection of two-body subsystems, as done for example in the 3C approximation. A sequence of two-body collisions within one or more of these subsystems leads to certain structures in the cross sections, which in the angular correlation appear under the so-called "critical angles" (Ref. [79] and references therein).

As we stated in Eq. (1.19), the scattering potential may consist of a number of scattering centers, e.g., the nucleus and a bound electron. Therefore, in general the T matrix elements are a sum of scattering amplitudes that may interfere with each other thus leading to striking effects in the cross sections. For example, in scattering from ground-state atomic helium it turned out that the scattering amplitudes from the two bound electrons and from the nucleus interfere such that for certain $\boldsymbol{k}_1$ and $\boldsymbol{k}_2$ the cross section is zero. This theoretical prediction [108, 119] is confirmed experimentally [120].

The pair emission from a variety of ground states has also been studied, e.g., from autoionizing states, pair emission accompanied with excitations (see, e.g., [121] and references therein) and from an orbitally oriented target (by laser pumping) (see, e.g., [122] and references therein). Here a wealth of new phenomena appear, however vestiges of the structure we discussed previously still underlie much of the physics even in this case.

Another topic of research in this field has been the pair emission in a laser background and how the angular and energy correlation is modified by the presence of the laser (see, e.g., [123] and references therein). Here the problem has to be approached within a time-dependent framework which will not be discussed here.

4 Electron–electron Interaction in Extended Systems

The correlated motion of the electrons in condensed matter results in striking phenomena such as superconductivity and magnetism [124, 125]. Coulomb interactions also have a strong influence on the electric properties of materials, e.g., electronic correlations may even inhibit conduction via delocalized electrons and thus determine whether a material is a metal or an insulator [126, 127]. Furthermore, some aspects of the electron–electron coupling in condensed matter are currently investigated in view of potential applications in quantum information devices (cf., e.g., [128] and references therein). On the other hand, as we briefly outlined in Section 1.5, according to the Landau theory of Fermi liquids [38] the interacting electronic systems can be mapped onto a system of weakly interacting quasiparticles which constitutes a great simplification from a computational and a conceptual point of view. Indeed, Landau's approach turned out to be very successful in describing a variety of properties of some materials. Screening in an electronic system plays a key role in this context: In the event that the interactions among the valence electrons are not screened sufficiently the quasi-particle concept becomes questionable. Indeed, in this case electronic correlations result in a variety of phenomena that cannot be explained within conventional quasiparticle pictures. Known examples are high-temperature superconductivity in cuprates [129], the colossal magnetoresistance in certain materials [130], and the Kondo impurity and the Kondo lattice systems [131].

Hence, it is valuable to consider ways and tools to investigate the details of the electron–electron interaction in the condensed phase, such as the range and the frequency and the momentum-transfer dependence of the electron–electron interaction. Before presenting the two-particle correlation spectroscopy as a tool to map out the features of the electron–electron interaction in matter it is instructive to outline and introduce a way to quantify formally the properties of the electron–electron interaction in extended electronic systems. A more detailed account can be found in standard textbooks on the subject (e.g. [124, 125, 132, 145–151]).

Electronic Correlation Mapping: From Finite to Extended Systems. Jamal Berakdar

ISBN: 3-527-40350-7

The aim here is to introduce some widely used quantities for the quantification of electronic correlation and to point out their relationships to the correlated two-particle spectroscopy.

4.1 Exchange and Correlation Hole

To describe the electronic correlation in an N particle system one utilizes the reduced two-particle density matrix $\gamma_2(x_1, x_2, x_1', x_2')$. γ_2 is expressible in terms of the exact N-body wave function Ψ as [145]

$$\gamma_2(x_1, x_2, x_1', x_2') = N(N-1) \int \Psi(x_1, x_2, x_3, \cdots, x_N) \\ \Psi^*(x_1', x_2', x_3, \cdots, x_N)\, \mathrm{d}x_3 \cdots \mathrm{d}x_N \tag{4.1}$$

Hence, from the antisymmetry of the fermionic wave function we deduce that

$$\gamma_2(x_1, x_2, x_1', x_2') = -\gamma_2(x_2, x_1, x_1', x_2') \tag{4.2}$$

The variables $x_j,\ j = 1 \cdots N$ label the spin (m_{sj}) and the position ($\boldsymbol{r}_j$) coordinates. The two-particle probability density derives from γ_2 in the limit

$$\begin{aligned} \rho_2(x_1, x_2) &= \gamma_2(x_1, x_2, x_1, x_2) \\ &= N(N-1) \int |\Psi(x_1, \cdots, x_N)|^2\, \mathrm{d}x_3 \cdots \mathrm{d}x_N \end{aligned} \tag{4.3}$$

For fermions we conclude that ρ_2 vanishes for $x_2 = x_1 = x$, i. e.

$$\rho_2(x, x) = 0 \tag{4.4}$$

For completely independent particles $\rho_2(x_1, x_2)$ is related to the single-particle density $\rho(\boldsymbol{r}_1) = N \int |\Psi(x_1, \cdots, x_N)|^2\, \mathrm{d}m_{s_1}\, \mathrm{d}x_2 \cdots \mathrm{d}x_N$ via the equality

$$\rho_2(x_1, x_2) = \rho(x_1) \frac{N-1}{N} \rho(x_2) \tag{4.5}$$

Thus, even for noninteracting (but overlapping) fermions the antisymmetry of Ψ leads to a correlation among the particles, the Fermi correlation. As a result the two-particle density possesses a hole, the Fermi hole in the region around $x_1 = x_2$ [134, 135]. The hole does not appear however for noninteracting electrons, if $m_{s_1} \neq m_{s_2}$. In the Hartree–Fock (HF) theory for example, this kind of electronic correlation, also called Fermi correlation, is exactly taken into account. However, for $m_{s_1} \neq m_{s_2}$ the HF theory yields

$$\rho_2^{(HF)}(\boldsymbol{r}_1, \boldsymbol{r}_2) = \rho(\boldsymbol{r}_1)\rho(\boldsymbol{r}_2) \tag{4.6}$$

which does not reflect the influence (on ρ_2) of another type of interparticle correlations that stem from the electrostatic Coulomb interactions [136].

To describe the different aspects of electronic correlations one usually introduces the so-called correlation factor $f(x_1, x_2)$ and the conditional probability density as

$$\begin{aligned} \rho_2(x_1,x_2) &= \rho(x_1)\rho(x_2)\left[1+f(x_1,x_2)\right] \\ \Omega(x_2,x_1) &= \frac{\rho_2(x_1,x_2)}{\rho(x_1)}; \qquad \int \Omega(x_2,x_1)\mathrm{d}x_2 = N-1 \end{aligned} \tag{4.7}$$

$\Omega(x_2, x_1)$ is a measure of the probability of finding one electron at the position $\boldsymbol{r}_2$ if another electron is present at $\boldsymbol{r}_1$. The difference between $\Omega(x_2, x_1)$ and $\rho(x_2)$ and a non-vanishing f are inherently due to correlations (Fermi correlation, Coulomb correlation and self-interaction).

The depression in the electronic density is described by the so-called exchange and correlation hole (h_{xc}) [124], which is defined as

$$h_{\mathrm{xc}}(x_1,x_2) = \frac{\rho_2\,(x_1,x_2)}{\rho(x_1)} - \rho(x_2) = \Omega(x_2,x_1) - \rho(x_2) = \rho(x_2)f(x_1,x_2) \tag{4.8}$$

Hence, we see from Eqs. (4.7) and (4.8) that

$$\int h_{\mathrm{xc}}(x_1,x_2)\,\mathrm{d}x_2 = -1 \tag{4.9}$$

meaning that the xc hole corresponds to a deficit of exactly one electron; an electron and its hole constitute therefore an entity with no net charge. Furthermore, to highlight effects due to exchange and Coulomb correlations it is customary to introduce the separation

$$h_{\mathrm{xc}}(x_1,x_2) = h_{\mathrm{x}}^{(m_{s1}=m_{s2})}(\boldsymbol{r}_1,\boldsymbol{r}_2) + h_{\mathrm{c}}^{(m_{s1}\neq m_{s2})}(\boldsymbol{r}_1,\boldsymbol{r}_2) \tag{4.10}$$

where h_{x} (h_{c}) is called the exchange (Coulomb) hole[1]. The separation between exchange and correlation effects is achieved by expressing the exchange hole in terms of a hole that depends

[1] In this work we are interested in mapping the angular and the energy dependence of the momentum-space two-particle density. Conventionally, one studies the role of h_{xc} in determining the energetics of the system. For example, in density functional theory [145, 147, 149], one defines the exchange-correlation energy $E_{\mathrm{xc}}[\rho(\boldsymbol{r})]$ using the method of adiabatic connection. In this method the noninteracting system with a fixed density is related to the interacting system via the coupling-constant λ, that allows the adiabatic switching of the electron–electron interaction. For $\lambda = 0$ we retrieve the noninteracting system and for $\lambda = 1$ the fully interacting system. $E_{\mathrm{xc}}[\rho(\boldsymbol{r})]$ is then expressed in terms of h_{xc} as

$$E_{\mathrm{xc}}[\rho(\boldsymbol{r})] = \frac{1}{2}\int \rho(\boldsymbol{r})\,\mathrm{d}^3\boldsymbol{r} \int \frac{h_{\mathrm{xc}}(\boldsymbol{r},\boldsymbol{r}')}{|\boldsymbol{r}-\boldsymbol{r}'|}\mathrm{d}\boldsymbol{r}' \tag{4.11}$$

Here the exchange and correlation hole $h_{\mathrm{xc}}(\boldsymbol{r},\boldsymbol{r}')$ is obtained by averaging over the coupling-constant dependent hole $h^{\lambda}_{\mathrm{xc}}(\boldsymbol{r},\boldsymbol{r}')$, i.e.,

$$h_{\mathrm{xc}}(\boldsymbol{r},\boldsymbol{r}') = \int_0^1 h^{\lambda}_{\mathrm{xc}}(\boldsymbol{r},\boldsymbol{r}')\,\mathrm{d}\lambda \tag{4.12}$$

on the electron–electron coupling constant[2], λ (cf. Eq. (4.12))

$$h_x(\boldsymbol{r}, \boldsymbol{r}') = h_{xc,\lambda=0}(\boldsymbol{r}, \boldsymbol{r}')$$

The correlation (Coulomb) hole is then given as

$$h_c(\boldsymbol{r}, \boldsymbol{r}') = h_{xc,\lambda}(\boldsymbol{r}, \boldsymbol{r}') - h_x(\boldsymbol{r}, \boldsymbol{r}')$$

From these relations we deduce the following sum rules

$$\int h_x(\boldsymbol{r}, \boldsymbol{r}')\,\mathrm{d}\boldsymbol{r}' = -1, \qquad h_x(\boldsymbol{r}, \boldsymbol{r}') \leq 0 \tag{4.16}$$

$$\int h_c(\boldsymbol{r}, \boldsymbol{r}')\,\mathrm{d}\boldsymbol{r}' = 0 \tag{4.17}$$

These stringent requirements set the spatial range of h_x. A deep h_x has a short spatial extension, and vice versa. On the other hand $h_c(\boldsymbol{r}, \boldsymbol{r}')$ admits positive or negative values. Furthermore, as deducible from Eq. (4.15), for $\boldsymbol{r} = \boldsymbol{r}'$ the exchange hole behaves as $h_x \to 0$ when the electrons have opposite spins. For equal spin states of the electron pair and for $\boldsymbol{r} = \boldsymbol{r}'$ we find $h_x(\boldsymbol{r}, \boldsymbol{r}) = -\rho(\boldsymbol{r})$.

4.2 Pair-correlation Function

The depletion of the average electron density around an individual electron can be described by means of the pair-correlation function g_{xc} which is defined as

$$g_{xc}(x_1, x_2) = h_{xc}(x_1, x_2)/\rho(x_2) + 1 = \rho_2(x_1, x_2)/[\rho(x_1)\rho(x_2)] \tag{4.18}$$

One introduces further the exchange-correlation energy per particle, i.e., the energy density, $\varepsilon_{xc}[\rho(\boldsymbol{r})]$ as

$$\varepsilon_{xc}[\rho(\boldsymbol{r})] = \frac{1}{2}\int \frac{h_{xc}(\boldsymbol{r}, \boldsymbol{r}')}{|\boldsymbol{r} - \boldsymbol{r}'|}\,\mathrm{d}\boldsymbol{r}' \tag{4.13}$$

These relations illustrate that the energetics of the electronic many-body problem is solved once $h_{xc}(\boldsymbol{r}, \boldsymbol{r}')$ is known exactly.

[2]Specifically, h_x is retrieved from the Hartree–Fock expression for the exchange energy,

$$E_x(\boldsymbol{r}, \boldsymbol{r}') = \frac{1}{2}\int \rho(\boldsymbol{r})\,\mathrm{d}\boldsymbol{r} \int \frac{h_x(\boldsymbol{r}, \boldsymbol{r}')}{|\boldsymbol{r} - \boldsymbol{r}'|}\,\mathrm{d}\boldsymbol{r}' \tag{4.14}$$

In terms of the spin orbitals, $\psi_j(\boldsymbol{r}m_s)$, h_x reads

$$h_x(\boldsymbol{r}, \boldsymbol{r}') = -\frac{1}{\rho(\boldsymbol{r})}\sum_{m_s}\left[\sum_j^N |\psi_j^*(\boldsymbol{r}m_s)\psi_j(\boldsymbol{r}'m_s)|\right]^2 \tag{4.15}$$

As we discussed, $h_{\rm xc}$ is in fact an average over the coupling constant λ. Consequently the pair-correlation function is given by $g_{\rm xc}(\boldsymbol{r},\boldsymbol{r}') = \int_0^1 g_{\rm xc}^{\lambda}(\boldsymbol{r},\boldsymbol{r}')\,{\rm d}\lambda$, where $g_{\rm xc}^{\lambda}$ corresponds to $h_{\rm xc}^{\lambda}$.

For an electron located at $\boldsymbol{r}$ the function $g_{\rm xc}(\boldsymbol{r},\boldsymbol{r}')$ yields the relative[3] probability of finding another electron at the point $\boldsymbol{r}'$ within an infinitesimal spatial volume ${\rm d}\boldsymbol{r}'$. In line with this interpretation, $g_{\rm xc}(\boldsymbol{r},\boldsymbol{r}')$ tends to unity at large r' and diminishes for $r' \to r$.

The general trends of the pair-correlation function (4.18) for varying relative strengths of the exchange or the correlation interactions are as follows:
If $h_{\rm x}$ dominates over $h_{\rm c}$ ($h_{\rm xc} \approx h_{\rm x}$) we infer the behavior of the pair-correlation function by noting that for equal spin states of the electron pair

$$g_{\rm xc}(\boldsymbol{r},\boldsymbol{r}) = 0|_{h_{\rm xc}\to h_{\rm x}}, \quad \text{and for inequal spin states} \quad g_{\rm xc}(\boldsymbol{r},\boldsymbol{r}) = 1|_{h_{\rm xc}\to h_{\rm x}}$$

so that the average value of the pair-correlation function is $g_{\rm xc}(\boldsymbol{r},\boldsymbol{r}) = 0.5|_{h_{\rm xc}\to h_{\rm x}}$ (cf. also the explicit example in Appendix D.2).

Including the influence of Coulomb correlations reduces $g_{\rm xc}(\boldsymbol{r},\boldsymbol{r})$ so that for spin-unpolarized electrons $g_{\rm xc}(\boldsymbol{r},\boldsymbol{r})$ is restricted to the interval

$$0 \le g_{\rm xc}(\boldsymbol{r},\boldsymbol{r}) \le \frac{1}{2} \tag{4.19}$$

In the case where the Coulomb correlations are dominant we find

$$g_{\rm xc}(\boldsymbol{r},\boldsymbol{r}) \to 0$$

From Eq. (4.18) it is evident that $g_{\rm xc}$ is a central quantity in the theory of correlated many-body systems, for it encompasses $h_{\rm xc}$. Over the past 60 years an impressive number of theoretical studies have been devoted to the investigation of the various aspects of the exchange-correlation hole in the condensed phase using a variety of theoretical techniques and approximations (see, e.g., Refs. [137–144] and references therein). Here we will not repeat or summarize these studies but rather concentrate on how one can actually measure $h_{\rm xc}(\boldsymbol{r}_1,\boldsymbol{r}_2)$ or $g_{\rm xc}(\boldsymbol{r}_1,\boldsymbol{r}_2)$ and how can we describe the experimental outcome theoretically.

[3] Relative means here that the two-particle probability density is normalized to the uncorrelated probabilities of two uncorrelated, classical electrons.

4.2.1 Effect of Exchange on the Two-particle Probability Density

For illustration let us consider the simplest example of a two-particle state $\Psi_{\boldsymbol{k}_1,\boldsymbol{k}_2}(\boldsymbol{r}_1, \boldsymbol{r}_2)$ consisting of two plane waves characterized by the two wave vectors $\boldsymbol{k}_1$ and $\boldsymbol{k}_2$ and consider the manifestation of exchange in the pair-correlation function, as given by Eq. (4.18). In the singlet channel g_{xc} is uniform. In the triplet channel we find that

$$g_{\mathrm{xc}}(\boldsymbol{r}_1, \boldsymbol{r}_2) = 1 - \cos[(\boldsymbol{k}_1 - \boldsymbol{k}_2) \cdot (\boldsymbol{r}_1 - \boldsymbol{r}_2)] \tag{4.20}$$

If the two electrons reside in a Fermi sea with N distinct occupied $\boldsymbol{k}$ states one readily deduce for the ground-state average of g_{xc} the expression

$$\langle g(\boldsymbol{r}_1, \boldsymbol{r}_2)\rangle = \frac{1}{N^2} \sum_{ij} \left(1 - \mathrm{e}^{\mathrm{i}(\boldsymbol{k}_i - \boldsymbol{k}_j)\cdot \boldsymbol{R}}\right) \tag{4.21}$$

$$= 1 - F^2(k_{\mathrm{F}} R) \tag{4.22}$$

where $\boldsymbol{R} = \boldsymbol{r}_2 - \boldsymbol{r}_1$, k_{F} is the Fermi wave vector. The function F has the form

$$F(k_{\mathrm{F}}R) = 3\frac{\sin(k_{\mathrm{F}}R) - (k_{\mathrm{F}}R)\cos(k_{\mathrm{F}}R)}{(k_{\mathrm{F}}R)^3} \tag{4.23}$$

These equations allow the following statements:

- g_{xc} tends to zero (the Fermi hole) for vanishing interparticle distances, even in the absence of Coulomb correlations. This is because we assumed the spin channels are resolved which demonstrates the importance of spin-resolved studies in disentangling Coulomb and Fermi correlations.

- Furthermore, since we assumed the system to be homogenous we find that

  $$\langle g(\boldsymbol{r}_1, \boldsymbol{r}_2)\rangle = \langle g(R)\rangle \tag{4.24}$$

  So in this case g depends on one variable only and can thus be accessed with single-particle probing techniques (cf. Appendix D for the measurable quantities).

- If the system is lattice periodic we can still express g_{xc} in terms of a single variable with a sum over the reciprocal lattice (cf. Appendix B).

- In general, however, and particularly for correlated disordered systems, $g_{\mathrm{xc}}(\boldsymbol{r}_1, \boldsymbol{r}_2)$ is most suitably investigated by means of two-particle spectroscopy which enables the study of local-field effects on the two-particle correlations.

- From Eq. (4.23) we see that averaging over the ground state results in damped oscillations and a finite range of the exchange-induced hole. This averaging becomes redundant if in an experiment (such as the one we shall discuss below) $\boldsymbol{k}_1$ and $\boldsymbol{k}_2$ are resolved.

4.3 Momentum-space Pair Density and Two-particle Spectroscopy

The aim of this section is to establish the relationships between the outcome of the experiment sketched in Fig. 2.1 and the pair-correlation function[4]. In principle, we have already encountered the formal equations (1.1) and (1.43) as recipes to evaluate the cross sections for the measurements depicted in Fig. 2.1. These equations are one variant of a more general expression that relies on the S matrix formalism which is explored here in some detail.

4.3.1 The S Matrix Elements

The probability P_{if} for the experiment shown in Fig. 2.1 is given by [3]

$$P_{\mathrm{if}} = S_{\mathrm{if}} S_{\mathrm{if}}^*, \tag{4.25}$$

where the S matrix elements are determined by

$$S_{\mathrm{if}} = \langle \Psi_{E_{\mathrm{f}}} | \Psi_{E_{\mathrm{i}}} \rangle \tag{4.26}$$

The normalized wave functions Ψ_{E_i} and Ψ_{E_f} describe the total system, i.e., the external test charge coupled to the sample with energies E_{i} and E_{f} in the initial and in the final state, respectively. Ψ_{E_i} and Ψ_{E_f} differ in their asymptotic behavior (boundary conditions). Ψ_{E_i} develops out of the uncorrelated asymptotic initial state $\phi_{E_{\mathrm{i}}}$. The asymptotic state $\phi_{E_{\mathrm{i}}}$ is a direct product of an undistorted ground state of the sample and a plane wave describing the incident electron. The final-channel asymptotic state $\phi_{E_{\mathrm{f}}}$ consists of the two free (detector) electron states and a sample that has lost one electron.

The mapping from $\phi_{E_{\mathrm{i/f}}}$ onto $\Psi_{E_{\mathrm{i/f}}}$ is mediated by the so-called Møller operator [152] which is defined in terms of a reference Green's function $G_0^{\pm}$ (satisfying appropriate boundary conditions) and an interaction potential $V_{\mathrm{i/f}}$, that is defined as $(G^{\pm})^{-1} - (G_{0,\mathrm{i/f}}^{\pm})^{-1} = V_{\mathrm{i/f}}$,

[4] The relationships to h_{xc}, Ω or the correlation factor f follow then from the respective definitions in Section 4.1.

as follows

$$
\begin{aligned}
\left|\Psi^{\pm}_{E_{\mathrm{i/f}}}\right\rangle &= |\phi_{\mathrm{i/f}}\rangle + G^{\pm}_{0,\mathrm{i/f}}(E_{\mathrm{i/f}})V_{\mathrm{i/f}}\,|\phi_{\mathrm{i/f}}\rangle + G^{\pm}_{0,\mathrm{i/f}}(E_{\mathrm{i/f}})V_{\mathrm{i/f}}G^{\pm}_{0,\mathrm{i/f}}(E_{\mathrm{i/f}})V_{\mathrm{i/f}}\,|\phi\rangle + \cdots \\
&= \left\{ \mathbb{1} + \left[G^{\pm}_{0,\mathrm{i/f}}(E_{\mathrm{i/f}}) + G^{\pm}_{0,\mathrm{i/f}}(E_{\mathrm{i/f}})V_{\mathrm{i/f}}G^{\pm}_{0,\mathrm{i/f}}(E_{\mathrm{i/f}}) + \cdots \right] V_{\mathrm{i/f}} \right\} |\phi_{\mathrm{i/f}}\rangle \\
&= \underbrace{\left[\mathbb{1} + G^{\pm}(E_{\mathrm{i/f}})V_{\mathrm{i/f}}\right]}_{\Omega^{\pm}_{\mathrm{i/f}}} |\phi\rangle \\
&= \overbrace{\left[\mathbb{1} + G^{\pm}_{0,\mathrm{i/f}}(E_{\mathrm{i/f}})T_{\mathrm{i/f}}\right]}\;|\phi_{\mathrm{i/f}}\rangle
\end{aligned}
\tag{4.27}
$$

The last equation exposes also the relationship between the Møller operator and the T operator which is defined by the Lippmann–Schwinger equation [153]

$$
T_{\mathrm{i/f}} = V_{\mathrm{i/f}} + V_{\mathrm{i/f}}G_{0,\mathrm{i/f}}T_{\mathrm{i/f}} \tag{4.28}
$$

The scattering operator (or S operator) is given by

$$
S = \Omega_f^{-\dagger}\Omega_i^{+} \tag{4.29}
$$

This relation allows us to express the transition probability (4.26) in terms of the asymptotic states as well as in terms of the density of states. The latter is deducible from the discontinuity of the Green's function at the branch cut, i.e. as a difference between the advanced and the retarded Green's function. The so-called post form of the S matrix elements reads [3]

$$
\begin{aligned}
S_{\mathrm{if}} = \langle \phi_{E_{\mathrm{f}}}|S|\phi_{E_{\mathrm{i}}}\rangle &= \left\langle \phi_{E_{\mathrm{f}}}|\Omega_f^{-\dagger}\Omega_i^{+}|\phi_{E_{\mathrm{i}}}\right\rangle \\
&= \langle \Psi^{-}_{E_{\mathrm{f}}}|\Psi^{+}_{E_{\mathrm{i}}}\rangle \\
&= \delta_{f,i} + \left\langle \phi_{E_{\mathrm{f}}}\middle| V_f\left[G^{+}(E_{\mathrm{f}}) - G^{-}(E_{\mathrm{f}})\right]\middle|\Psi^{+}_{E_{\mathrm{i}}}\right\rangle \\
&= \delta_{f,i} - \mathrm{i}2\pi\delta(E_{\mathrm{f}} - E_{\mathrm{i}})\,\langle \phi_{E_{\mathrm{f}}}|V_f|\Psi^{+}_{E_{\mathrm{i}}}\rangle
\end{aligned}
\tag{4.30}
$$

Similarly one can express the S matrix elements in terms of the asymptotic initial state and the exact final state in terms of the equivalent relation (prior form of the S matrix elements)

$$
S_{\mathrm{fi}} = \delta_{f,i} - \mathrm{i}2\pi\delta(E_{\mathrm{f}} - E_{\mathrm{i}})\,\langle \Psi^{-}_{E_{\mathrm{f}}}|V_i|\phi_{E_{\mathrm{i}}}\rangle \tag{4.31}
$$

Equations (4.25), (4.31), (4.26) and (4.30) are the basis for the physical interpretation of the two-particle coincidence experiments:

1. If the state $\langle \Psi^-_{E_\mathrm{f}}|$ is known one can measure P_if and deduce from Eqs. (4.26) the magnitudes of the projections of the state $\langle \Psi^-_{E_\mathrm{i}}|$ onto the known state $\langle \Psi^-_{E_\mathrm{f}}|$.

2. From Eqs (4.31) and (4.30) one can also deduce information on the type of the frequency and momentum-dependent potential $V_{\mathrm{i/f}}$ that acts between a test charge (the incoming particle in Fig. 2.1) and the sample, i.e., in the linear response regime one obtains detailed information on the charge susceptibility of the sample (see Appendix B.2 for a brief introduction).

4.3.2 Transition Probabilities and Cross Sections

To establish the connection between Eqs. (4.25), (4.26), (4.31) and (4.30) and the relation (1.3) that we discussed in the previous chapter one uses a formulation in the time domain by employing the identity

$$2\pi\delta(E_\mathrm{f}-E_\mathrm{i}) = \int_{-\infty}^{\infty} \mathrm{d}t\ \mathrm{e}^{\mathrm{i}(E_\mathrm{f}-E_\mathrm{i})t}$$

The time-dependence of the transition probability can be then deduced from the on-shell part of the matrix elements T_fi of the transition operator (in the prior or in the post formulation) or by evaluating the expression $P_\mathrm{if} = S_\mathrm{fi}S^*_\mathrm{fi}$. Conventionally one then introduces a transition rate, i. e. a transition probability per unit time as

$$\frac{\mathrm{d}P_\mathrm{fi}}{\mathrm{d}t} = 2\delta_{f,i}\Im T_\mathrm{fi} + 2\pi\delta(E_\mathrm{f}-E_\mathrm{i})|T_\mathrm{fi}|^2 \tag{4.32}$$

Therefore, for inelastic transitions the equation applies

$$\frac{\mathrm{d}P_\mathrm{fi}}{\mathrm{d}t} = 2\pi\delta(E_\mathrm{f}-E_\mathrm{i})\ T_\mathrm{fi}|^2. \tag{4.33}$$

Considering now the situation of Fig. 2.1, one usually defines a cross section as the transition rate per momentum-space interval $\mathrm{d}\mathcal{K} \equiv \mathrm{d}^3\boldsymbol{k}_1\mathrm{d}^3\boldsymbol{k}_2\mathrm{d}^3\boldsymbol{k}\mathrm{d}^3\boldsymbol{g}$ normalized to the incoming asymptotic probability flux density j_p, i.e., we write

$$\sigma(\mathcal{K}) := \frac{\mathrm{d}P_\mathrm{fi}}{\mathrm{d}t}\frac{1}{j_\mathrm{p}}\mathrm{d}\mathcal{K} \tag{4.34}$$

$$= (2\pi)^4\frac{1}{v_0}|T_\mathrm{fi}|^2\delta(E_\mathrm{f}-E_\mathrm{i})\mathrm{d}\mathcal{K} \tag{4.35}$$

which is the relation (1.3) we employed in previous chapters as a basis for the calculations and interpretations. Usually, the reciprocal wave vectors $\boldsymbol{g}$ and the Bloch wave vectors $\boldsymbol{k}$ (crystal momentum) are unknown and one has to average and respectively sum over these quantities.

4.3.3 Two-particle Emission and the Pair-correlation Function

In some cases a frozen-core picture for the electron-pair excitation is viable, i.e. we can assume that only the degrees of freedom of an electron pair with the initial and final energies E_i and E_f are affected while the surrounding medium is frozen. For example, in the case shown in Fig. 2.1 the frozen-core approximation is most suitable when the electron emitted from the sample originates from the top of the valence band. In this case we may write that

$$\Psi_{E_\mathrm{i}} \approx \psi_\mathrm{i}(x_1, x_2)\chi_M(x_3, \cdots, x_N) \tag{4.36}$$

ψ_i is the initial-state wave function of the interacting electron pair that consists of one electron in the valence band of the sample and the approaching test charge. The medium surrounding this electron pair is described by χ_M. Within this decoupling the reduced density matrix (4.1) attains the form

$$\gamma_2\left(x_1, x_2, x_1', x_2'\right) \approx 2\psi\left(x_1, x_2\right)\psi^*\left(x_1', x_2'\right) \tag{4.37}$$

A particularly transparent form of the transition probability is obtained if we assume further that the final-state electron pair is described by plane waves. In this case the measured, spin (m_{sj}) unresolved probability reads

$$P_\mathrm{if}(\boldsymbol{k}_1, \boldsymbol{k}_2) \propto \sum_{m_{s_1}, m_{s_2}, m_{s_1'}, m_{s_2'}} \tilde{\psi}(m_{s_1}\boldsymbol{k}_1, m_{s_2}\boldsymbol{k}_2)\tilde{\psi}^*(m_{s_1'}\boldsymbol{k}_1, m_{s_2'}\boldsymbol{k}_2) \tag{4.38}$$

Here $\tilde{\psi}$ is the double Fourier transform of ψ. Hence, in the case where (4.38) is applicable, the transition probability measured by the experiment depicted in Fig. 2.1 is proportional to the spin-averaged diagonal elements of the reduced density matrix in momentum space, i.e. the spin-averaged momentum-space two-particle density ρ_2. Since the momentum-space single particle densities can also by measured using the same apparatus shown in Fig. 2.1 (by switching off one of the detectors) one can normalize ρ_2 to the single particle densities and obtains thus the momentum-space pair-correlation function. The relation between $P_\mathrm{if}(\boldsymbol{k}_1, \boldsymbol{k}_2)$ and ρ_2 means that all properties of ρ_2 are reflected in P_if, in particular $P_\mathrm{if}(\boldsymbol{k}_1, \boldsymbol{k}_2)$ vanishes for $\boldsymbol{k}_1 = \boldsymbol{k}_2$. Furthermore, as checked theoretically (cf. Chapter 7) and experimentally [154] $P_\mathrm{if}(\boldsymbol{k}_1, \boldsymbol{k}_2)$ will in general not be a function of merely one argument $\boldsymbol{k}_1 - \boldsymbol{k}_2$, in contrast to Eq. (4.24) which is valid in the case of (exchange) coupled free particles.

As evident from Fig. 2.1 an energy and angular-resolved experiment determines by means of energy and momentum conservation law the value of $\boldsymbol{k}_1$ and $\boldsymbol{k}_2$ so that the averaging of the pair-correlation function over the initial state can, in principle, be reduced to an interval set by the energy and angular resolution of the experiment. This is important in so far as such an average changes qualitatively the form and the range of g_{xc}, as demonstrated in the preceding section. Furthermore, as mentioned above, the spin resolution holds the promise of studying the influence of exchange effects. Such highly demanding experiments have already been conducted [155–157], even though a complete mapping of the spin-resolved h_{xc} in momentum space is still outstanding.

In this context it should be mentioned that, from a theoretical point of view, the distinction between initial and final state when describing the experiment shown in Fig. 2.1 is rather formal. As evidenced by Eqs. (4.31) and (4.30) we can equally calculated P_{fi} using correlated initial or final states. The reason for this is the assumption that the detection time is much larger than the time scales of all processes in the systems which may have an effect on the pair emission. This results in a standing flux scattering formulation of the pair emission and in an equivalence between forward and backward time propagations (prior and post form). Nevertheless we can test experimentally for the influence of (physical) final and initial configuration of the system. For example, in Fig. 2.1 if one measures $P_{\mathrm{fi}}(\boldsymbol{k}_1, \boldsymbol{k}_2, \boldsymbol{k}_0)$ one can test for the initial state by leaving $\boldsymbol{k}_1$ and $\boldsymbol{k}_2$ fixed and varying $\boldsymbol{k}_0$. Since the final state depends mainly on $\boldsymbol{k}_1$ and $\boldsymbol{k}_2$ one may then assign the changes in P_{fi} to changes in the initial state.

5 The Electron–Electron Interaction in Large Molecules and Clusters

The research efforts in dealing with electronic correlations in condensed matter systems are paralleled with equally intensive research on the role of electronic correlations in molecular and polymeric materials (cf. for instance [124, 158–161] and references therein). In this chapter we are interested in a particular aspect of the electron–electron interaction in finite systems, namely the relationship between the wave-vector and the frequency-dependent electron–electron interaction and the system size and the number of electrons. The aim is to explore the potential of the correlated two-particle emission process for studying these aspects of the electron–electron interaction. In particular we will consider the fullerene molecule C_{60} and metal clusters. There are a number of indications of the importance of the electron–electron interaction in determining the physical properties of C_{60}. For example, it has been discovered that C_{60} when doped with alkali metals becomes superconductive [162]. The KVV Auger studies [163] on C_{60} indicate that doped C_{60} is indeed a strongly correlated system, for instance it was argued that K_3C_{60} is a half-filled Mott–Hubbard insulator. Furthermore, it was concluded theoretically that for solid (ordered) phase C_{60} the screened on-site molecular Coulomb integral is as large as $\approx$ 2.1 eV [164]. The issue of concern here is how the electron–electron interaction is modified by screening. This aspect has been addressed for solid C_{60}, e.g., [165, 166]. In this chapter we will follow the traces in the two-particle emission spectrum of the dynamical and the wave-vector dependent screening in free (gas-phase) molecules and clusters. It is noteworthy to mention that various spectroscopic studies, such as photoemission, inverse photoemission, and soft X-ray absorption spectroscopy [167]), have established the similarities of the electronic structure of C_{60} in the gas and in the solid phase which makes the conclusions drawn in this chapter of relevance to surface-deposited fullerenes. Two-particle emission intensities from ordered C_{60} have been also determined experimentally [168]. An extensive survey of the properties of fullerenes can be found in numerous textbooks, e.g., [169–171]

Electronic Correlation Mapping: From Finite to Extended Systems. Jamal Berakdar

ISBN: 3-527-40350-7

In the gas phase in a series of experiments [32–35, 173] the probability for the removal of one electron from the valence band of the carbon fullerenes (C_{60}) has been measured by bombarding it with electrons. Density functional calculations (DFT) within the local-density approximation (LDA) as well as Hartree–Fock calculations failed to reproduce the excitation probabilities as a function of the excitation energy [36, 172, 174]. The missing important element in these calculations is that the electron–electron interaction[1] is treated as a local, instantaneous perturbation, neglecting thus dynamical screening effects due to the presence of the surrounding medium. As outlined in Appendix B, using the random-phase approximation (RPA) such phenomena can be treated within a linear response approach (i.e., a linear response of the valence-band charge cloud to the field of the approaching test charge). Thus, a step forward to go beyond the DFT-LDA or the HF calculations for the electron-removal probability is to perform the calculations using RPA. Such an undertaking makes sense if screening is substantial, for RPA is essentially a first-order perturbation approximation in the screened electron–electron interaction. Performing such a task within DFT-LDA is computationally demanding and has not yet been performed. Here we present and discuss a feasible implementation of a self-consistent Hartree–Fock procedure. The numerical realization of this scheme has been performed with the the so-called variable-phase method which is capable of dealing with a large number of electrons. In addition, we perform the RPA procedure appropriate to electron removal processes. The RPA we employ here is similar in spirit to the conventional one presented in Appendix B.2. For calculational efficiency reasons as well as for illustration we develop and apply the RPA procedure in this chapter on the basis of the HF theory.

5.1 Retardation and Nonlocality of the Electron–Electron Interaction in Extended Systems

Let us consider the electron dynamics of an electronic quantum system under the influence of an external time-dependent perturbation U. Specifically, we will study below the electron-removal upon charged-particle impact, i.e., the same process as depicted in Fig. 2.1. The perturbation U is then the interaction between the incoming "test" electron and the ground-

[1] More precisely, the electron–electron interaction meant here is that between the approaching projectile electron and the valence-band electron.

state electronic charge distribution of the target. The question of interest here is how the electron–electron interaction is affected by the presence of a surrounding polarizable medium and how properties of the resulting effective interaction U_{eff} are manifested in the electron-removal probabilities from the valence shell. In Appendix (B) we address this issue within the Kubo formalism, here we briefly show how U_{eff} can be treated within a wave-function based approach.

The quantum dynamics of the system is governed by the Hamiltonian

$$\widehat{H}(t) = \widehat{H}_0 + U(t) \tag{5.1}$$

In what follows $\widehat{H}_0$ is assumed to be the self-consistent mean-field Hamiltonian that describes the ground state of the system. The solution $\Psi(\boldsymbol{r}, t)$ of the time-dependent Schrödinger equation $\left[\mathrm{i}\partial_t + \widehat{H}(\boldsymbol{r}, t)\right] \Psi(\boldsymbol{r}, t) = 0$ can be expressed as an antisymmetrized product of the single-electron orbitals $\psi_i(\boldsymbol{r}, t)$ with an associated energy ϵ_i as follows

$$\Psi(\boldsymbol{r}, t) = \mathrm{e}^{-\mathrm{i}E_0 t} \det \|\psi_i(\boldsymbol{r}, t)\| \tag{5.2}$$

The ground-state Hartree–Fock energy is denoted by E_0 which is evaluated as the following expectation value ($|i\rangle$ is a shorthand notation for $|\psi_i(\boldsymbol{r}, t)\rangle$)

$$E_0 = \sum_i \langle i| - \frac{\nabla}{2} - V_{\text{ions}} |i\rangle + \frac{1}{2} \sum_{i,k} \langle ik| \, u \, |ik - ki\rangle \tag{5.3}$$

Here $u = \frac{1}{|\boldsymbol{r} - \boldsymbol{r}'|}$ is the naked, instantaneous electron–electron interaction. V_{ions} is the ionic potential. Expanding the time-dependent, single-particle orbitals $\psi_i(\boldsymbol{r}, t)$ on a basis of time-independent Hartree–Fock orbitals we can write

$$\psi_i(\boldsymbol{r}, t) = N_i \left[\varphi_i(\boldsymbol{r}) + \sum_m C_{mi}(t) \varphi_m(\boldsymbol{r}) \right] \tag{5.4}$$

Here the sum runs over discrete and continuum states. The particle states, i.e., states which are energetically higher than the Fermi level E_{F}, are referred to by m. The hole states are indexed in this equation by i. The factor N_i is deduced from the normalization requirement. The expansion coefficients $C_{mi}(t)$ in Eq. (5.4) yield the probability amplitudes for the creation of the m–i electron–hole pair. Now we insert Eq. (5.4) into Eq. (5.2) and impose that

$$\langle \Psi(\boldsymbol{r}, t)| \, \widehat{H} - i\frac{\partial}{\partial t} \, |\Psi(\boldsymbol{r}, t)\rangle \equiv 0 \tag{5.5}$$

A first-order determining equation for the particle–hole excitation amplitudes $C_{mi}(t) \neq 0$ follows then from the relation[2]

$$
\begin{aligned}
i \sum_{i \le E_F < m} C^*_{mi}(t) \frac{\partial}{\partial t} C_{mi}(t) \quad = \quad & \sum_{i \le E_F < m} \Bigg\{ (\varepsilon_m - \varepsilon_i) \left| C_{mi}(t) \right|^2 \\
+ \quad & C_{mi}(t) \langle i | U | m \rangle + C^*_{mi}(t) \langle m | U | i \rangle \\
+ \quad & \sum_{j \le E_F < k} \Bigg[\frac{1}{2} C^*_{mi}(t) C^*_{kj}(t) \langle mk | u | ij - ji \rangle \\
+ \quad & \frac{1}{2} C_{mi}(t) C_{kj}(t) \langle ij | u | mk - km \rangle \\
+ \quad & C^*_{mi}(t) C_{kj}(t) \langle mi | u | kj - kj \rangle \Bigg] \Bigg\}
\end{aligned}
\tag{5.7}
$$

Hence, we obtain via variation the following relation for $C^*_{mi}(t)$

$$
\begin{aligned}
i \frac{\partial}{\partial t} C_{mi}(t) = & \left(\varepsilon_m - \varepsilon_i\right) C_{mi}(t) + \langle m | U | i \rangle \\
& + \sum_{j \le E_F < k} \left[C^*_{kj}(t) \langle mk | u | ij - ij \rangle + C_{kj}(t) \langle mj | u | ik - ki \rangle \right]
\end{aligned}
\tag{5.8}
$$

To solve this equation we make the following ansatz

$$
C_{mi}(t) = X_{mi} \mathrm{e}^{-\mathrm{i} E_q t} + Y^*_{mi} \mathrm{e}^{\mathrm{i} E_q t} \tag{5.9}
$$

E_q is the energy transferred to the system upon the passage of the incident projectile electron. The ansatz (5.9) when inserted in Eq. (5.8) leads to the following two coupled equations for the coefficients[3] X_{mi} and Y^*_{mi}

$$
\begin{aligned}
& \left(\varepsilon_m - \varepsilon_i - E_q\right) X_{mi} + \langle m | U | i \rangle \\
& \quad + \sum_{j \le E_F < k} \left[\langle mj | u | ki - ik \rangle X_{kj} + \langle mk | u | ji - ij \rangle Y_{kj} \right] = 0
\end{aligned}
\tag{5.10}
$$

[2]The linear terms in $C_{mi}(t)$ that do not contain the external, time-dependent field cancel out as solutions of the Hartree–Fock equation

$$
\langle m | - \frac{\nabla}{2} - V_{\text{ions}} | i \rangle + \sum_{j \le E_F} \langle mj | u | ij \rangle = 0 \tag{5.6}
$$

[3]Note that if the time-independent part of the external perturbation and the inter-electronic interaction u vanish the functions X and Y become independent solutions that describe respectively the electron excitation and de-excitation from the i to the m level.

$$(\varepsilon_m - \varepsilon_i + E_q)\, Y_{mi} + \langle i|\, U \,|m\rangle + \sum_{j \le E_F < k} \left[\langle ij|\, u \,|km - mk\rangle\, X_{kj} + \langle ik|\, u \,|jm - mj\rangle\, Y_{kj} \right] = 0 \quad (5.11)$$

These two equations can be formulated in terms of the matrix elements of the effective operator U_{eff} by means of the following definitions

$$\langle m|\, U_{\text{eff}} \,|i\rangle = -(\varepsilon_m - \varepsilon_i - E_q) X_{mi} \quad (5.12)$$

$$\langle i|\, U_{\text{eff}} \,|m\rangle = -(\varepsilon_m - \varepsilon_i + E_q) Y_{mi} \quad (5.13)$$

This leads to the determining integral equation for the effective interaction U_{eff}

$$\langle m|\, U_{\text{eff}} \,|i\rangle = \langle m|\, U \,|i\rangle + \sum_{j \le E_F < k} \left[\frac{\langle k|\, U_{\text{eff}} \,|j\rangle \, \langle mj|\, u \,|ki - ik\rangle}{E_q - \varepsilon_k + \varepsilon_j + i\eta} + \frac{\langle j|\, U_{\text{eff}} \,|k\rangle \, \langle mk|\, u \,|ji - ij\rangle}{E_q + \varepsilon_k - \varepsilon_j - i\eta} \right] \quad (5.14)$$

Here $\eta \ll 1$ is a positive real number. The integral equation (5.14) which is also called the random-phase approximation with exchange (RPAE) dictates how the action of U is modified by the response of the system and expresses this modification in terms of the dynamical function U_{eff}. In general U_{eff} is energy dependent and nonlocal. Comparing Eq. (5.14) with Eqs. (C.1) and (C.2) we see that U_{eff} has the same structure as derived within the Kubo formula, namely $U_{\text{eff}} = \epsilon^{-1} U$ where ϵ plays the role of the dielectric response function of the sample.

We recall that within the HF approximation exchange effects are taken exactly into account but Coulomb correlations, in the sense detailed in the preceding chapter, are not accounted for (cf. Eq. 4.6). In contrast, within the local-density approximation of the density functional theory (DFT-LDA) exchange and correlation effects are described in an approximate way through the employed exchange and correlation functional. The advantage of the HF of incorporating exchange effects exactly (and being self-interaction free) comes at the expense of evaluating the expectation value of a large number of nonlocal potentials (Fock terms) [36, 175]. In the examples shown below the numerical problems arising from the nonlocality of the potentials have been circumvented by utilizing a nonlocal version of the variable-phase method (VPM) [176].

5.2 Electron Emission from Fullerenes and Clusters

Appendix B.2 outlines briefly how the effective interaction $U_{\mathrm{eff}}(\boldsymbol{q}, \omega = E_q/\hbar)$ in an extended system is modified by the various excitation modes of the sample, such as the particle–hole pair and plasmon creations. In particular we mentioned the decisive influence of the dimensionality. Here we address the question of how the size of the system and the number of active (valence) electrons are manifested in the behavior of the effective interaction $U_{\mathrm{eff}}(\boldsymbol{q}, \omega = E_q/\hbar)$ and how the frequency and the momentum-dependent response can be tested by means of two-particle spectroscopy.

Let us consider the situation where a test charge approaches with a momentum $\boldsymbol{k}_0$ a cluster of atoms in the ground state. The effective interaction U_{eff} between the test charge and the sample results in the emission of a valence electron. The scattered test charge and the emitted electron emerge in the final channel with momenta $\boldsymbol{k}_1$ and $\boldsymbol{k}_2$. According to Eq. (5.14) the transition amplitude for this reaction is

$$T(\boldsymbol{k}_0, \varphi_\nu; \boldsymbol{k}_1, \boldsymbol{k}_2) = \langle\, \boldsymbol{k}_1 \boldsymbol{k}_2 | U_{\mathrm{eff}} | \varphi_\nu \boldsymbol{k}_0 \rangle \tag{5.15}$$

where φ_ν is a single particle orbital describing an electron in the valence band with energy ε_ν. The matrix elements of U_{eff} are determined by solving for the integral equations (5.14).

If electrons are employed as a test charge one can test for the spin-dependent effective interaction in the way detailed in Chapter 2. Unfortunately such experiments are not available yet and the theory has to average over the spin degrees of freedom. Nevertheless, we will show below how the influence of exchange on the effective interaction can be assessed even in the spin-unresolved measurements. In the case of an electron as a test charge, the spin-averaged angular and energy-resolved cross section $\sigma(\boldsymbol{k}_0; \boldsymbol{k}_1, \boldsymbol{k}_2)$ is given as a weighted average of the singlet $\propto |T^{(S=0)}(\boldsymbol{k}_0, \varphi_\nu; \boldsymbol{k}_1, \boldsymbol{k}_2)|^2$, i.e., vanishing total spin ($S = 0$) of the final-state electron pair, and the triplet $\propto |T^{(S=1)}(\boldsymbol{k}_0, \varphi_\nu; \boldsymbol{k}_1, \boldsymbol{k}_2)|^2$ cross sections (here we consider exchange effects only). Experimental data are available for the total cross section $\sigma(E_0)$ which is calculated as

$$\sigma(E_0) = \frac{(2\pi)^4}{k_0} \int \mathrm{d}^3\boldsymbol{k}_1 \mathrm{d}^3\boldsymbol{k}_2 \left\{ \sum_\nu \frac{1}{4} \left| T^{(S=0)}(\boldsymbol{k}_0, \varphi_\nu; \boldsymbol{k}_1, \boldsymbol{k}_2) \right|^2 \right.$$
$$\left. + \frac{3}{4} \left| T^{(S=1)}(\boldsymbol{k}_0, \varphi_\nu; \boldsymbol{k}_1, \boldsymbol{k}_2) \right|^2 \right\} \delta\left(E_0 + \varepsilon_\nu - (k_1^2/2 + k_2^2/2)\right) \tag{5.16}$$

In the event that one employs as a test charge a projectile other than an electron (e.g., protons or positrons) the exchange contribution to the cross section vanishes, however other channels might need to be included (e.g., the capture of the target electron by the projectile).

As is clear from Eqs. (5.15), (5.16) and (5.14), in order to calculate the cross section the valence band and the scattering states of the cluster are needed. For a practical implementation one can calculate the single particle and hole states of the sample using a self-consistent Hartree–Fock procedure. In this way one incorporates in the calculations of the sample's ground state the mean-field part of the electron–electron interactions and the exchange contributions exactly. In the examples shown below the bound and the scattering wave functions are computed simultaneously by utilizing the nonlocal variable phase method [176].

5.2.1 The Spherical Jellium Model

In the case of C_{60} the spherical jellium model [177, 178] has proved useful. In this approach the atomic potentials experienced by the valence-band electrons are modelled by a shell confinement, i.e., the radial motion of the valence electrons is confined by the potential well:

$$V(r) = V_0 \quad \text{for} \quad R - \Delta < r < R + \Delta, \quad \text{and} \quad V = 0 \text{ elsewhere} \tag{5.17}$$

Here r is the radial distance measured from the center of the fullerene. For C_{60} the following values are used: $R \approx 6.7\,a_0$ which is the radius of the fullerene. The average C–C bond length enters in the model potential as the thickness of the shell (2Δ 2.69 a.u.). The depth of the potential V_0 is set by the (experimental) first ionization potential of C_{60} which is 7.6 eV and by the number of valence-band electrons which is 240. Figure 5.1 illustrates the type of the radial self-consistent field potential $V_{SC}(r)$ for the valence electrons. For the purpose of demonstration the calculated local part only is shown for a valence band electron with a zero angular momentum. Figure 5.1 evidences that the mean-field part of the electron–electron interaction and the local part of the exchange interaction modify markedly the external confining potential (5.17).

As demonstrated in [172, 174], the substantial extent of the valence-band orbitals of C_{60} and the large number of relevant electrons (240 e^-) cause severe convergence problems when calculating the transition matrix elements (even without performing the RPAE loop). Hence, the development and implementation of efficient numerical procedures are indispensable [176].

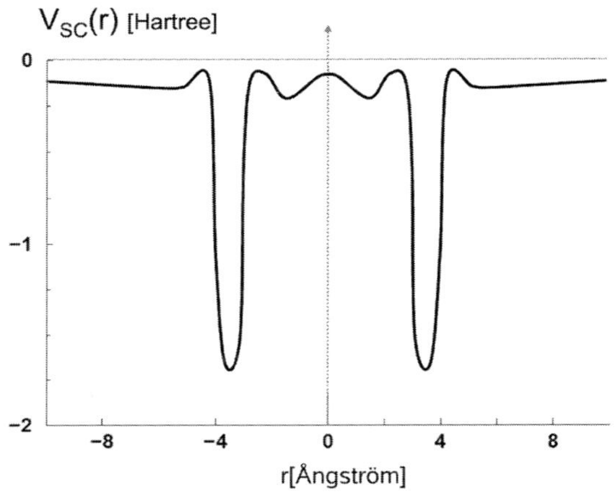

Figure 5.1: The calculated local part of the radial self-consistent single-particle potential $V_{SC}(r)$ of C_{60} with a zero angular momentum. $r = 0$ is the center of the fullerene cage.

5.2.2 Angular Pair Correlation

Figure 5.2 shows the angular dependence of the fully differential cross section $\sigma(\boldsymbol{k}_0; \boldsymbol{k}_1, \boldsymbol{k}_2)$ for the emission of an electron from the highest occupied molecular orbital (HOMO) of C_{60} upon the impact of 50 eV electrons. The calculations have been performed with (upper part in Fig. 5.2) and without (lower part in Fig. 5.2) solving numerically for the RPAE equation (5.14).

As demonstrated schematically in Fig. 5.2, the amount of momentum transfer $\boldsymbol{q} = \boldsymbol{k}_0 - \boldsymbol{k}_1$ can be controlled depending on the chosen emission angle θ_1 and the energy E_1 of the projectile. In this way one tunes to the limit of far distance collision (small q) where the Thomas–Fermi (TF) approximation is justifiable (cf. Appendix C.1). The TF model is advantageous in so far as it offers a qualitative interpretation of the results in the way demonstrated below. On the other hand increasing q one enters the close collision regime and the TF approach to screening is no longer viable. So the present technique allows, in principle, the study of the crossover from the long to the short-wavelength behavior of screening. Thereby the two-particle nature of the effective electron–electron interaction is accessible, as demonstrated in Fig. 5.2. One important limitation of the two-particle coincidence technique is however the following: The projectile has to transfer to the target at least the amount of the single ionization energy for the HOMO electron to be ejected. This implies a minimal momentum transfer

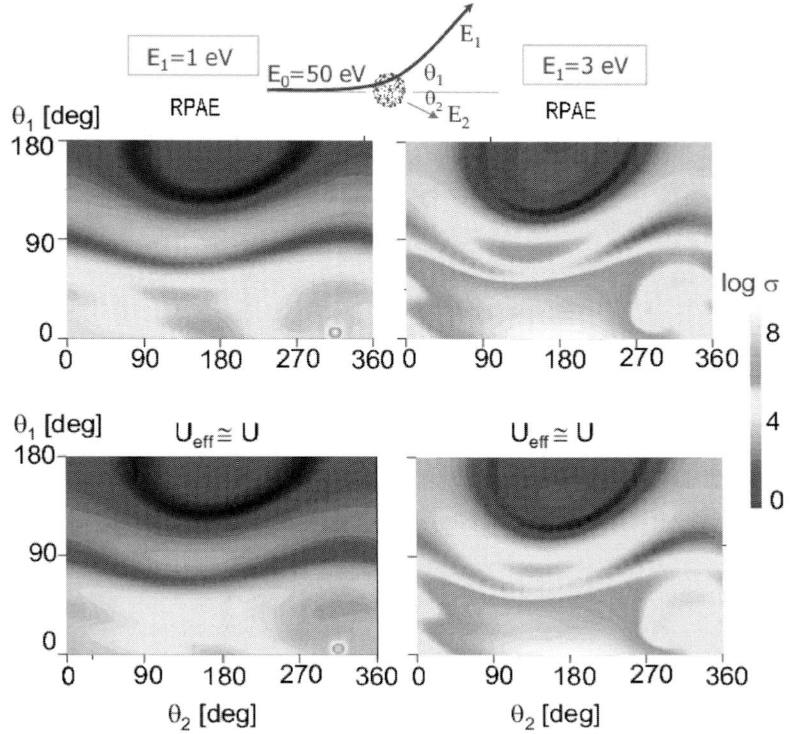

Figure 5.2: The angular dependence of the fully differential cross section for the emission of one electron from C_{60} with an energy of 1 eV (left panel) or 3 eV (right panel) following the impact of 50 eV electrons. A schematic of the scattering geometry is depicted. The upper part of the figure shows the RPAE calculations while the lower part shows the calculations without treating screening effects. See also color figure on page 149.

and hence the excitation modes of the sample that occurs with zero-momentum transfer cannot be investigated, e.g., as we discuss in Appendix C.3 in the three-dimensional case the plasma mode is most suitably studied in the limit of zero-momentum transfer to avoid decay into particle–hole pairs.

In Fig. 5.2 the momentum transfer q is kept relatively small so that U_{eff} can be qualitatively modelled by the TF potential $U_{\text{eff}}(r) \sim \left[\mathrm{e}^{-r/\lambda}\right]/r$ where λ is the screening length (which in our finite-system case is a parameter). Furthermore, since the potential we are using is almost constant at large distances (cf. Fig. 5.1) and the electrons in the geometry of Fig. 5.2 have energies well above threshold we can argue that the scattering states are qualitatively

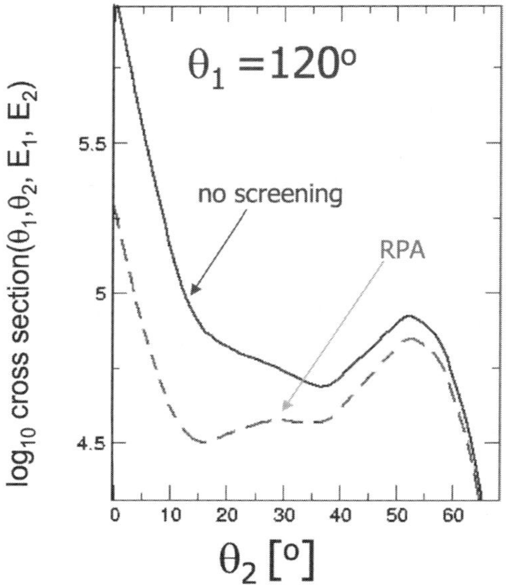

Figure 5.3: The same scattering geometry as in Fig. 5.2, however the electron that has the energy 1 eV is detected under the fixed angular position $\theta_1 = 120^\circ$ whereas the angular distribution of the other electron with the energy 3 eV is scanned. The results of the RPAE calculations (dashed line) are shown along with those done while neglecting screening effects (solid line).

modelled by plane waves (this has also been checked to be true numerically). Under these assumptions the cross section has then qualitatively the same structure as given by Eq. (1.7) which we discussed in Chapter 1, i.e., it is determined by the form factor $\tilde{U}_{\text{eff}}$ of U_{eff} where $\tilde{U}_{\text{eff}}(q) \propto (q^2 + 1/\lambda^2)^{-1}$ and by the momentum-space wave function of the valence-band electron.

The influence of a substantial screening, i.e., small λ as compared to $1/q$ is apparent from the form of $\tilde{U}_{\text{eff}}$: if $\lambda \ll 1/q$ the form factor tends to $\tilde{U}_{\text{eff}}(q) \sim \lambda^2$ and hence we deduce from Eq. (1.7) that, in this case, the cross section is generally reduced and the binary and recoil peaks are smeared out. Structures occurring in the angular distributions are then assigned to the structure of the momentum-space representation of the initially bound state (cf. Eq. (1.7)). This tendency is already observed in Fig. 5.3 where we see the suppression of the cross section when screening effects are incorporated. Using the same reasoning similar behavior can be deduced for the energy correlation between the electrons [182]. Unfortunately, in both cases there are no experiments yet available.

5.2.3 Total Cross Sections

The spin-unresolved total cross section, as given by Eq. (5.16), has been measured [32, 33]. The experimental data and the corresponding calculations are depicted in Fig. 5.4. In this figure we also included the results of the calculations using density functional theory (DFT) within the local-density approximation (LDA) [174] where a similar model potential for the cluster as given in Eq. (5.17) has been employed. As is clear from Fig. 5.4 the DFT-LDA model predicts essentially the same qualitative behavior for the cross section as does the present theory without screening.

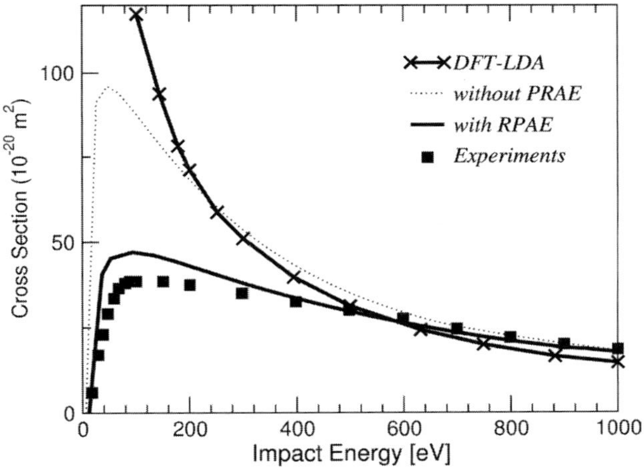

Figure 5.4: The absolute total cross section for the removal of one electron from C_{60} upon the inelastic collision of electrons with the impact energy displayed on the axis. The experimental data (full squares) are taken from Refs. [32, 33]. The solid curve with crosses is the result of DFT calculations [172] whereas the dotted curve shows the present theory without accounting for the RPAE corrections. Theoretical results based on RPAE are shown by the solid curve.

Two notable features are observed in Fig. 5.4: In the high (impact) energy regime we notice hardly any influence of charge-density response, i.e., calculations with and without screening are almost equivalent. This behavior is comprehensible as the time scale for the retarded response of the target does not match the very short passage time of keV electrons. In this situation the process can be viewed as in the atomic case, i.e., the incoming electron interacts with the HOMO electron while the surrounding charge cloud is frozen on the time scale of the interaction and emission times. In contrast, an incorporation of the effect of the charge-density response on the electron–electron interaction is indispensable at lower excitation energies.

Another point worth noting is that the shape of the measured and the calculated RPAE cross sections show a saturation behavior instead of the sharply peaked shape in the atomic or unscreened case. This behavior becomes comprehensible when modelling the screening using the TF model: as we analyzed in full details in Chapter 1 and demonstrated in Fig. 1.3 the total cross section shows a saturation behavior with decreasing screening length λ. On the other hand one can reverse the argument and ask the question which screening length is needed in the TF potential in order to achieve the behavior depicted in Fig. 5.4. This issue has been addressed in [36] with the result that $1/\lambda \sim 0.3$ a.u. is the appropriate value. This substantially strong screening justifies in retrospect the use of RPAE (first-order perturbation approximation in the screened electron–electron interaction). In this context it should be mentioned that while the TF potential contains one single constant parameter λ this does not mean that the screening effect on the cross section will be energy- or angle-independent. This is because the potential U_{eff} enters as a part of the six-dimensional integrand in Eq. (5.15) and hence the cross section is not simply related to U_{eff}. On the other hand Fig. 5.4 demonstrates that screening of the electron–electron interaction is inappropriately modelled by a constant suppression of the Coulomb interaction (U/ϵ_0), where ϵ_0 is a dielectric constant, for such an approximation does not change the shape of the cross section but leads to a mere energy-independent lowering of the cross section.

5.2.4 Finite-size Effects

As stated above one of our goals is the investigation of the influence of the system size on the two-particle interaction. Generally, from the structure of Eq. (5.14), we can conclude that if the system size shrinks the level spacing grows and if the number of electrons is small the contributions of the particle–hole excitations and de-excitations (i.e., the sum in Eq. (5.14)) become generally irrelevant compared to the contributions of the matrix elements of the naked interaction (the first term in Eq. (5.14)). This general anticipation is confirmed by the total cross-section calculations within the spherical jellium model for the the electron-removal from Li clusters of varying sizes. The cross sections depicted in Fig. 5.5 are normalized to the number of electrons in the respective clusters. As is clear from this figure, with decreasing size of the cluster the influence of the particle–hole (de-)excitations on the cross section diminishes, i.e., the cross section results of the calculations with and without RPAE are almost identical. In contrast, when the system size increases the influence of screening as triggered by the charge-density response modifies the cross section in a characteristic way (cf. Fig. 5.4), particularly at low energies. This leads to the counterintuitive phenomena that the normalized cross section decreases with increasing cluster size because screening is more effective for larger clusters.

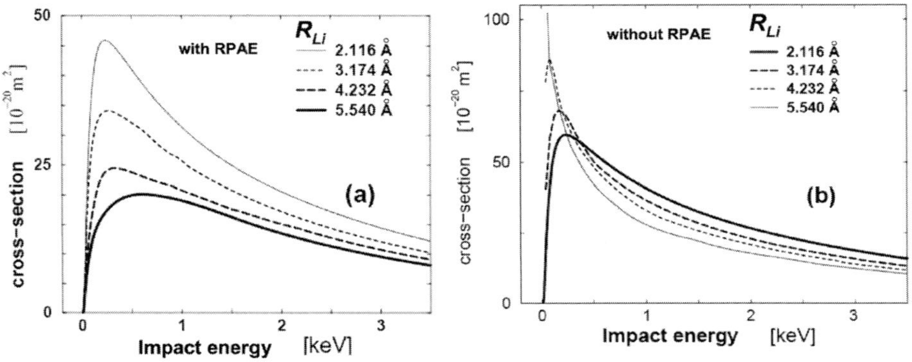

Figure 5.5: The electron-impact ionization cross section for spherical Li clusters with varying radius size R_{Li}. (a) shows the RPAE calculations. (b) shows the results when the particle–hole (de)excitations are neglected.

As in the case of Fig. 5.4, the cross sections for large clusters also show saturation behavior as a function of the impact electron energy. For small clusters we recover the shape of the cross section akin to atomic systems, i.e., a sharp peak at around a few times the ionization potential. The physics behind this behavior was discussed in Chapter 1 (cf. Fig. 1.3). The findings presented in Figs. 1.3, 5.4 and 5.5 indicate that generally the shape of the total charged-particle impact ionization cross section as a function of the impact energy is dictated by the charge-density response of the target. Indeed, this observation is also confirmed by a number of experiments on large and biological molecules [37].

5.2.5 Influence of Exchange

In principle, a detailed investigation of exchange, and spin-dependent effects in general, entails the use of spin-polarized projectiles (cf. Chapter 2) which is experimentally a much more demanding task. Nevertheless, an indication of the influence of exchange effects can also be deduced from spin-averaged cross sections, e.g., by utilizing the symmetry properties of

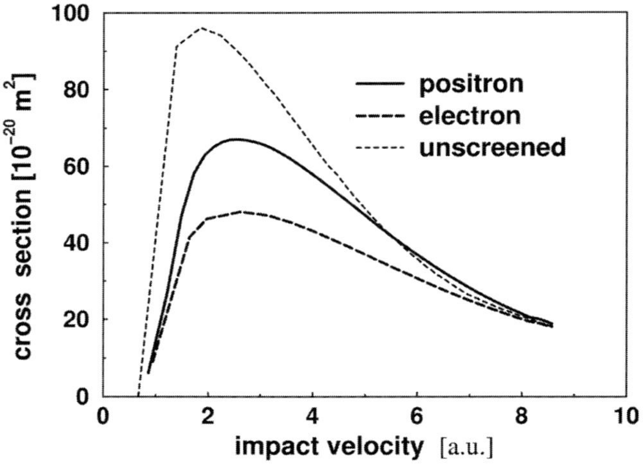

Figure 5.6: Electron and positron-impact ionization cross sections of HOMO of C_{60}. Solid curve shows the result of the RPA calculations for the case of positron impact. Dashed curve is the RPAE cross sections for the electron impact including exchange. The dotted curve stands for the calculations without account for screening. These latter cross sections are (within the present model) the same for electron and positron impact.

the singlet and triplet cross section, as demonstrated in some detail in Chapter 3. Here we illustrate a further method to get some insight in the role of exchange effects: By comparing electron and positron impact cross sections from the HOMO of the C_{60} molecule, as done in Fig. 5.6. The total cross sections are calculated according to Eq. (5.14). However, in the case of the positron the exchange contribution to the screening is missing[4]. Within our model the first term of Eq. (5.14), i.e., neglecting the dynamical screening altogether, results in cross sections which are independent of the charge state of the projectile, i.e. these cross sections are the same for positron and electron impact.

Generally, accounting for the influence of the change-density response reduces the cross section for the reasons detailed above. At high incidence velocities both the electron and the positron projectiles lead to the same total (unscreened) cross section. The reason for the diminishing influence of exchange effects in this limit is that, at higher energies, fast collisions with small q contribute predominantly to the cross sections (cf. Chapter 1) and hence the fast incoming projectile also emerges as a high velocity. This means, in the case of a projectile electron, the the two final-state electrons become effectively distinguishable

[4]This amounts to the use of the so-called Tamm–Dancoff approximation [146].

via their substantially different energies and hence the exchange effects play no role in this case. At low energies however, exchange effects play a decisive role. Unfortunately, there are as yet no experiments on C_{60} using positrons as a projectile. On the other hand numerous experiments have been conducted on the ionization channel of C_{60} using protons as projectiles ([180, 181] and references therein). Here, particularly in the low velocity regime, one has to account for the channel where one electron of C_{60} is captured by the projectile in addition to the direct ejection of one electron from the target into a detector state. These channels can be discriminated in a fully resolved experiment, i.e., in the case where the charge state of the projectile is also resolved in the final channel.

Another point worth further consideration in this context is the influence of the charge-density response related to the inter-cluster interaction in solid C_{60} and how this is manifested in the two-particle spectrum. Such experiments have already been conducted on ordered C_{60} deposited on a single copper crystal [179]. An appropriate theoretical model is a tight-binding approach, as outlined in Appendix A, since the deposited clusters are well-separated from each other and the experiments indicate relatively small interaction of the cluster with the substrate [179].

6 Pair Emission from Solids at Surfaces

Experiments on the correlated electron-pair emission from thin solid films were conducted [183] shortly after their atomic counterpart [72, 73]. It is however, only recently that these surface-science experiments have been performed in a full-resolved manner and with a sufficient accuracy [5, 6, 67, 154, 157, 184–187]. The main experimental obstacle is the low two-particle coincidence rate as compared to the overwhelming background of secondary electrons. This entails a long detection time while maintaining the stability of the same experimental conditions.

On the other hand, a quantum theoretical treatment of the pair-emission at surfaces is equally demanding. It has to incorporate the following ingredients: (i) One requires the characteristic response of the sample as well as the renormalized two-particle interaction at the particular energy and momentum transfer chosen by the experiment. (ii) The propagation of the incoming and the outgoing particles has to be described in the combined potential composed of the crystal potential and the effective two-particle potential. (iii) It is further necessary to develop or utilize a reasonable model for the electronic structure of the sample. A number of further elements may have to be incorporated in the theory as they can decisively influence the pair-emission intensity; for instance we mention here disorder effects, surface roughness, scattering from impurities, and spin-related phenomena such as spin–orbit effects. Some of these facets, which we shall address below, have been included successfully in current theoretical treatments [68, 188–192].

A number of aspects of the correlated pair emission are also encountered in the theory of conventional spectroscopies such as EELS, SPEELS and angular resolved photoemission spectroscopy (cf., e.g., [18–20, 40, 42] and references therein). There one also has to address the problem of how to describe single-electron propagation in solids or at surfaces and how to account for possible losses of energy and wave vectors of the electrons. An approach which

Electronic Correlation Mapping: From Finite to Extended Systems. Jamal Berakdar

ISBN: 3-527-40350-7

proved successful in this regard is to describe the sample's ground-state properties within a single-particle picture relying on the concept of DFT-LDA. Using self-consistent methods such as (linear muffin tin orbitals [193] or the full-potential linearized augmented-plane wave method [194, 195]) one then solves for an effective one-particle Schrödinger equation. The quasiparticle character of the excited states, as outlined in Section 1.5, is approximately captured by the use of a complex self-energy in addition to the single-particle potentials, i.e., in effect one can still operate within a picture of a single-particle moving in an effective field [196, 197]. The results of this approach for the LEED, SPLEED and the photoemission spectra are in quantitative agreement with experiment[1] [198,199,202,214]. This methodology is also retained here for the description of the single-particle states.

The present chapter starts with the consideration of ordered materials using a description of the single-particle states in terms of Bloch waves. To demonstrate the generalities of some of the conclusions drawn in this chapter we investigate the case where the projectile is a point charge carrying the charge $Z_{\rm p}$ and the mass $m_{\rm p}$ (for example, electrons, positrons, protons, ...). Quantities corresponding to the incoming projectile such as the wave vector, the energies and the scattering angles, are therefore indexed by p. When electron impact is studied we retain the notation introduced in Fig. 2.1.

6.1 Qualitative Analysis

The electron emission following the excitation of the surface by the incoming projectile is described by the transition matrix elements that are expressible as the sum of two amplitudes

$$\mathcal{T} = T_{\rm pe} + T_{\rm pc}. \tag{6.1}$$

For the interpretation of the results, particularly for electron impact, it is instructive to transform canonically to the wave vector representation $\boldsymbol{K}^- \otimes \boldsymbol{K}^+$ where $\boldsymbol{K}^+ = \boldsymbol{k}_1 + \boldsymbol{k}_2$ is the center-of-mass wave vector of the two-particle system. $\boldsymbol{K}^- = (\boldsymbol{k}_1 - \boldsymbol{k}_2)/2$ characterizes the interparticle wave vector (i.e., $\boldsymbol{K}^-$ describes the internal degree of freedom of the two-particle system viewed as a single composite particle whose location is determined by $\boldsymbol{K}^+$).

[1] It should be mentioned however, that the self-energy is usually not derived in an ab initio way but in many cases is parametrized so as to fit experimental data. This issue will be addressed in more detail in the next chapter.

The direct pair emission amplitude $T_{\rm pe}$ arises from the interaction of the incoming test charge with the emitted electron via the effective interaction $U_{\rm eff}$. It has the form

$$T_{\rm pe} = \langle \boldsymbol{K}^-, \boldsymbol{K}^+ | U_{\rm eff} | \boldsymbol{k}_0, \varphi_{\epsilon_i(\boldsymbol{k})} \rangle \tag{6.2}$$

The initial state $|\boldsymbol{k}_0, \varphi_{\epsilon_i(\boldsymbol{k})}\rangle$ describe the two-particle states consisting of the incoming particle with a wave vector $\boldsymbol{k}_0$ and a conduction band electron with the wave function $\varphi_{\epsilon_i(\boldsymbol{k})}$. ϵ_i is the single particle energy and $\boldsymbol{k}$ is the corresponding Bloch wave vector.

In the geometry depicted in Fig. 2.1, i.e., in the back-reflection geometry, a scattering from the crystal is vital. In fact due to the (linear) momentum conservation law, a classical scattering in the back-reflection geometry cannot lead to the emission of two particles in the backward direction. The scattering of the interacting two-particle system from the (semi-infinite) crystal potential $W_{\rm pc}$ is described by the transition amplitude $T_{\rm pc}$ which has the form [68]

$$T_{\rm pc} = \iint \mathrm{d}^3\boldsymbol{p}\, \mathrm{d}^3\boldsymbol{q} \langle \boldsymbol{K}^-, \boldsymbol{K}^+ | U_{\rm eff} g^-_{\rm pe} | \boldsymbol{p}, \boldsymbol{q} \rangle \langle \boldsymbol{p} | W_{\rm pc} | \boldsymbol{k}_0 \rangle \langle \boldsymbol{q} | \varphi_{\epsilon_i(\boldsymbol{k})} \rangle \tag{6.3}$$

In this equation $|\boldsymbol{q}\rangle \otimes |\boldsymbol{p}\rangle$ is a complete set of plane waves. $W_{\rm pc}$ is the interaction potential between the projectile and the lattice. The quantity $g^-_{\rm pe}$ stands for the electron–projectile Green's function (with outgoing wave boundary conditions). Hence the operator $U_{\rm eff} g^-_{\rm pe}$ describes the propagation of the two-particle excited state $|\boldsymbol{p}, \boldsymbol{q}\rangle$ to the final detector states $\langle \boldsymbol{K}^-, \boldsymbol{K}^+|$.

6.1.1 Model Crystal Potential

For a more explicit form of the expression (6.3) a model for the crystal potential is needed. Let us consider metals with a dense sphere packing with a potential $V^{\rm ion}(r)$ around the ionic sites that has approximately a spherically symmetric shape and otherwise a smooth structureless form. Under these circumstances one can expand $W_{\rm pc}$ in a sum of spherical, non-overlapping potentials centered around the ionic sites (non-overlapping muffin-tin (MT) potentials [194, 215]). In the interstial region the potential is assumed constant and this constant potential region is usually chosen as the zero reference point (called muffin-tin zero). In the constant region the propagating particle waves are undistorted. Hence one has to solve the single site scattering problem and then construct the scattering amplitude from the muffin-tin potentials. In [200] the role of nonspherical contributions within the muffin-tin sphere has been investigated.

To a give a concrete example of how the MT potential is constructed and how the two-particle scattering is described let us consider a (semi-infinite) solid with a general orthorhombic Bravais lattice. The potential $W_{\rm pc}$ is assumed as a superposition of effective core potentials of the ions

$$W_{\rm pc} = \sum_{i}^{N} V_i^{\rm ion} \tag{6.4}$$

$V_i^{\rm ion}(r)$ is the spherically symmetric potential centered around the ionic site i and N is the number of ions. The energies of the electrons we are going to consider are low to intermediate (below 1 keV) so that, due to the short escape depth of the electrons, we have to deal essentially with a surface as a target, i.e., we have to consider the three-dimensional symmetry break due to the presence of the surface (a semi-infinite solid). If the sample is viewed as a collection of ℓ atomic layers then the potential $W_{\rm pc}$ is periodic only in each layer parallel to the $x - y$ plane (cf. Fig. 2.1). The lattice constants in the x, y directions are d_x, d_y. Due to possible reconstruction near the surface d_z may vary near the surface/vacuum interface. However, away from the surface $z \ll 0$ the three-dimensional periodicity is restored with a lattice constant d_z, meaning that d_z is a function of z. The jth ion in the ℓth layer has the coordinates $\boldsymbol{r}_{j,\ell} = (\boldsymbol{r}_{\parallel,j}, r_{\perp,\ell})$. Thus, the periodic potential $W_{\rm pc}$ at the position $\boldsymbol{r}'$ can be written as

$$W_{\rm pc}(\boldsymbol{r}'_{\parallel}, z') = \sum_{\ell} \sum_{j} V^{\rm ion}(\boldsymbol{r}'_{\parallel} - \boldsymbol{r}_{\parallel,j}, z' - r_{\perp,\ell}) \tag{6.5}$$

Because $W_{\rm pc}$ is periodic in the x and y directions it is useful to operate in the reciprocal space spanned by the two-dimensional reciprocal vectors $\boldsymbol{g}_{\parallel} = 2\pi(n_x/d_x, n_y/d_y),\ n_x, n_y \in \mathbb{Z}$. The potential $W_{\rm pc}$ at the position $\boldsymbol{r}'$ is then expressed as

$$W_{\rm pc}(\boldsymbol{r}'_{\parallel}, z') = \sum_{\boldsymbol{g}_{\parallel}} \tilde{W}_{\rm pc}(\boldsymbol{g}_{\parallel}, z') \exp(\mathrm{i}\boldsymbol{g}_{\parallel} \cdot \boldsymbol{r}'_{\parallel}) \tag{6.6}$$

The two-dimensional Fourier transform $\tilde{W}_{\rm pc}(\boldsymbol{g}_{\parallel}, z')$ is cast as

$$\tilde{W}_{\rm pc}(\boldsymbol{g}_{\parallel}, z') = \frac{1}{A_{\rm uc}} \int_{A_{\rm uc}} d^2\boldsymbol{r}'_{\parallel} W_{\rm pc}(\boldsymbol{r}'_{\parallel}, z') \exp(-\mathrm{i}\boldsymbol{g}_{\parallel} \cdot \boldsymbol{r}'_{\parallel}) \tag{6.7}$$

The surface of the two-dimensional (2D) unit cell in the $x - y$ layer is denoted by $A_{\rm uc}$. The 2D integral in Eq. (6.7) is restricted to $A_{\rm uc}$.

6.1.2 Scattering from the Surface Potential

As is clear from Eq. (6.3) the (elastic) single-particle scattering amplitude from the potential (6.7) is required as an essential ingredient of the two-particle scattering amplitude T_{pc}. Namely, one has to evaluate the expression $\langle \boldsymbol{q}|W_{\mathrm{pc}}|\boldsymbol{k}_i\rangle$

$$\begin{aligned}\langle \boldsymbol{q}|W_{\mathrm{pc}}|\boldsymbol{k}_i\rangle &= (2\pi)^{-3}\sum_{\boldsymbol{g}_\parallel}\int \mathrm{d}^3\boldsymbol{r}_\mathrm{p}\exp(-\mathrm{i}\boldsymbol{K}\cdot\boldsymbol{r}_\mathrm{p}+\mathrm{i}\boldsymbol{g}_\parallel\cdot\boldsymbol{r}_{p,\parallel})\tilde{W}_{\mathrm{pc}}(\boldsymbol{g}_\parallel,z_\mathrm{p}),\\ &= (2\pi)^{-2}\sum_{\boldsymbol{g}_\parallel}\delta(\boldsymbol{g}_\parallel-\boldsymbol{K}_\parallel)\int \mathrm{d}z_\mathrm{p}\tilde{W}_{\mathrm{pc}}(\boldsymbol{g}_\parallel,z_\mathrm{p})\exp(-\mathrm{i}K_z z_\mathrm{p}) \qquad (6.8)\end{aligned}$$

In this equation the intermediate momentum transfer vector is given by $\boldsymbol{K} := \boldsymbol{q} - \boldsymbol{k}_i$, whereas $\boldsymbol{r}_\mathrm{p}$ refers to the position of the incoming projectile and $z_\mathrm{p} = \boldsymbol{r}_\mathrm{p}\cdot\hat{\boldsymbol{z}}$. The 2D lattice-translational symmetry thus results in the restriction that scattering is allowed only if $\boldsymbol{g}_\parallel = -\boldsymbol{K}_\parallel$ (Bragg condition). Substituting Eqs. (6.5) and (6.7) into Eq. (6.8) and noting that $N=\sum_j\exp(-\mathrm{i}\boldsymbol{g}_\parallel\cdot\boldsymbol{r}_{\parallel,j})$ we deduce the following relation for the Fourier transform $\tilde{W}_{\mathrm{pc}}(\boldsymbol{K})$

$$\begin{aligned}\tilde{W}_{\mathrm{pc}}(\boldsymbol{K}) &= \frac{N(2\pi)^2}{A_{\mathrm{uc}}}\sum_{\ell,\boldsymbol{g}_\parallel}\delta(\boldsymbol{g}_\parallel-\boldsymbol{K}_\parallel)\mathrm{e}^{-\mathrm{i}K_z r_{\perp,\ell}}\\ &\times \int_{-\infty}^{\infty}\mathrm{d}\boldsymbol{r}_z\int_{A_{\mathrm{uc}}} d^2\boldsymbol{r}_\parallel\, V^{\mathrm{ion}}(\mathbf{r}_\parallel,z_\mathrm{p})\mathrm{e}^{-\mathrm{i}(\boldsymbol{g}_\parallel\cdot\mathbf{r}_\parallel+K_z\boldsymbol{r}_z)} \qquad (6.9)\end{aligned}$$

Here we introduced the variable $\mathbf{r} = \boldsymbol{r}_\mathrm{p} - \boldsymbol{r}_j$. Since we are operating within a non-overlapping MT approximation the integration in Eq. (6.9) over the unit cell can be extended to cover the entire $x-y$ plane and we infer that

$$\tilde{W}_{\mathrm{pc}}(\boldsymbol{K}) = \frac{N(2\pi)^2}{A_{\mathrm{uc}}}\sum_{\ell}\mathrm{e}^{-\mathrm{i}K_z r_{\perp,\ell}}\sum_{\boldsymbol{g}_\parallel}\delta(\boldsymbol{g}_\parallel-\boldsymbol{K}_\parallel)\tilde{V}^{\mathrm{ion}}(\boldsymbol{K}) \qquad (6.10)$$

where $\tilde{V}^{\mathrm{ion}}(\boldsymbol{K})$ is the Fourier transform of V^{ion}.

So what is needed for the numerical calculations is an expression for V^{ion}. This can be approximated by a model potential for a qualitative study or can be calculated self-consistently using DFT-LDA. For some of the numerical examples shown below V^{ion} is deduced from DFT-LDA calculation and parametrized in the spherical form

$$V^{\mathrm{ion}}(\boldsymbol{r}_\mathrm{p}) = \frac{Z_{\mathrm{eff}}\mathrm{e}^{-\lambda_{\mathrm{eff}}r_\mathrm{p}}}{r_\mathrm{p}} \qquad (6.11)$$

Here the parameters Z_{eff}, λ_{eff} depend on the sample under study. A further simple, but rather unrealistic, approximation is to assume V^{ion} in the form of a contact (delta function) potential with some scattering length as a parameter. In this case analytical results can be obtained that can be utilized for a qualitative analysis.

6.1.3 Qualitative Features of Interacting Two-particle Emission from Surfaces

As we have seen above, the two-dimensional periodicity of the crystal potential results in the von-Laue diffraction condition, expressed by the delta function in Eq. (6.10). This result is in fact valid in general for ordered clean semi-infinite solids, regardless of the actual functional form of the on-site potential. Now making use of the Bloch theorem for the states of the scattered particles (cf. Appendix A) one shows the following. Assuming U_{eff} to be dependent on $\boldsymbol{r}_{\mathrm{p}} - \boldsymbol{r}_2$, the general form of the transition amplitude T_{pc} is

$$T_{\mathrm{pc}} = C \sum_{\ell, \boldsymbol{g}_{\|}} \delta[\boldsymbol{g}_{\|} - (\boldsymbol{K}_{\|}^{+} - \boldsymbol{K}_{0,\|})] \, \mathcal{L}(\boldsymbol{g}_{\|}, \ell, \boldsymbol{K}^{+}, \boldsymbol{K}^{-}, \boldsymbol{k}) \tag{6.12}$$

We conclude that T_{pe} is given by

$$T_{\mathrm{pe}} = \delta(\boldsymbol{K}_{0,\|} - \boldsymbol{K}_{\|}^{+}) \, \mathcal{L}' \tag{6.13}$$

In these equations we introduced the vector $\boldsymbol{K}_0 = \boldsymbol{k}_0 + \boldsymbol{k}$ which is the initial wave vector of the excited two particles. The functions $C, \mathcal{L}, \mathcal{L}'$ depend on the actual form of the states of the involved particles, a concrete example is given below. Here it is worthwhile to note the following conclusions based on the general form of Eqs. (6.12) and (6.13):

1. The scattering amplitude (6.12) contains the von-Laue diffraction condition for the center-of-mass wave vector of the two-particle subsystem. This is because T_{pc} involves not only the scattering from the crystal potential but also, at the same time, an energy and momentum exchange within the two-particles mediated by U_{eff}.

   This exchange occurs while conserving energy and momentum[2] which results in the pair-diffraction condition. We recall that in the single-electron scattering from the crystal potential (cf. Eqs. (6.8) and (6.10)), as occurs in LEED, diffraction occurs when the change in the wave vector of the incident electron matches $\boldsymbol{g}_{\|}$ [201, 202]. The difference here is that a given two-particle center-of-mass wave vector $\boldsymbol{K}^{+}$ does not imply necessarily that

[2] This is an approximation which we will address below.

the wave vectors $\boldsymbol{k}_1$ and $\boldsymbol{k}_2$ are fixed. The internal wave vector $\boldsymbol{K}^-$ is independent of $\boldsymbol{K}^+$ and can change even if $\boldsymbol{K}^+$ is fixed. The relevance of $\boldsymbol{K}^-$ to the pair diffraction is outlined below. The point here is that pair diffraction may occur when no diffraction is possible for the independent single particles and vice versa.

2. The change in the center-of-mass wave vector determines the *positions* of the diffraction peaks. On the other hand the function $\mathcal{L}$ which depends on $\boldsymbol{K}^-$, i.e., on the parameter characterizing the interparticle correlation, governs the *shape and intensity* of the diffraction peaks. Hence we have the situation that, in contrast to single-particle diffraction, say LEED, two-particle diffraction is truly a manifestation of the two-particle correlation within the scattered pair. In addition, it signals the coherent two-particle scattering from the crystal potential.

3. Usually the Bloch wave vector $\boldsymbol{k}_{\|}$ of the initially bound Bloch electron is not known, so one has to integrate the $\boldsymbol{k}$ resolved cross section over the Brillouin zone (BZ). As deduced from Eq. (6.12) this procedure results in a smearing of the two-particle diffraction pattern, even if $\boldsymbol{K}^+$ and $\boldsymbol{k}_0$ are accurately determined experimentally.

4. The last statement implies that if $\boldsymbol{K}_{\|}^+, \mathbf{g}_{\|}$ and $\boldsymbol{k}_{0,\|}$ are well resolved, the positions in the wave-vector space and the widths of the diffraction peaks reflect the behavior of the wave-vector distribution of the initial single-particle state of the sample.

6.1.4 Explicit Results for Two-particle Scattering from Metal Surfaces

To calculate the amplitude (6.12) numerically one requires, in addition to the crystal potential an expression for the Bloch states in the sample. For s–p bonded metals in which the ionic sites scatter the electrons only weakly one may approximate the ionic site potentials (i.e., the MT potential) by a constant positive 'background charge'. In this way the many-electron problem is reduced to the iterative self-consistent solution of a spin-independent one-body problem. In the jellium model[3], the resultant effective one-particle potential is replaced by a local step potential V_0 at $z = 0$. Within the metal half space $z < 0$ the electrons are confined within the

[3]This approximation serves only the calculations of the wave functions of the conduction-band electrons. It should be stressed that the discrete MT lattice potential still enters the calculations through Eq. (6.3). Only the sample's electronic state $\varphi_{\boldsymbol{k}}$ is obtained from the jellium model.

volume V by the potential barrier

$$V_0 = E_\mathrm{F} + W \tag{6.14}$$

where E_F is the Fermi energy and W is the work function. The density of states is given by that of the free- electron gas (and a factor 2 for the electronic spin states) $\rho_j = V/(4\pi^3)$. Within the jellium model we can thus write the spin-independent stationary states of the conduction band electrons in terms of reflection and transmission coefficients:

$$\langle \boldsymbol{r}_2|\varphi_{\boldsymbol{k}}\rangle = \frac{1}{\sqrt{V}} \exp(\mathrm{i}\mathbf{k}\cdot\boldsymbol{r}_{2,\parallel}) \begin{cases} \mathrm{e}^{\mathrm{i}\mathbf{k}_z z_2} + R\mathrm{e}^{-\mathrm{i}\mathbf{k}_z z_2} & z_2 < 0 \\ T\mathrm{e}^{-\gamma z_2} & z_2 > 0 \end{cases} \tag{6.15}$$

The reflection and transmission coefficients R and T are given by

$$R = \frac{\mathbf{k}_z - \mathrm{i}\gamma}{\mathbf{k}_z + \mathrm{i}\gamma}, \quad T = \frac{2\mathbf{k}_z}{\mathbf{k}_z + \mathrm{i}\gamma} \tag{6.16}$$

and $\gamma = \sqrt{2V_0 - \mathbf{k}_z^2}$.

Now we need an expression for the excited two-particle state in the vacuum. One may utilize LEED states for this purpose (as done below), however the influence of diffraction is already incorporated if the crystal potential (6.10) is used. For this reason, to obtain analytical results, the final scattering state is treated within the plane wave approximation, i.e.

$$\Psi_f(\boldsymbol{r}_2, \boldsymbol{r}_\mathrm{p}) = (2\pi)^{-3} \exp\left[i\boldsymbol{k}_\mathrm{p}\cdot\boldsymbol{r}_\mathrm{p} + \mathrm{i}\boldsymbol{k}_2\cdot\boldsymbol{r}_2\right] \tag{6.17}$$

It is possible to incorporate in this expression distortion effects due to image–charge interaction [68] at the expense of more complicated expressions for the scattering amplitudes. However, as it turned out [68], in the cases discussed below image–charge distortion effects do not alter qualitatively the behavior of the cross section.

Assuming the MT approximation is valid and the single-site potential is parameterizable in the form (6.11) and within the Thomas–Fermi approximation (C.7) to the projectile–electron interaction one finds [68] for the amplitudes T_pc the analytical expression

$$T_\mathrm{pc} = \sqrt{\frac{8}{\pi V}\frac{Z_\mathrm{p} Z_\mathrm{eff} N}{A_\mathrm{uc}}} \sum_{\ell, \boldsymbol{g}_\parallel} \delta[\boldsymbol{g}_\parallel - (\boldsymbol{K}_\parallel^+ - \boldsymbol{K}_{0,\parallel})]\,(\mathcal{L}_0 + \mathcal{L}_1 + \mathcal{L}_2) \tag{6.18}$$

where the functions $\mathcal{L}_j,\ j = 0, 1, 2$ are given by the following expressions

$$\begin{aligned} \mathcal{L}_0 \;&=\; \exp(-\mathrm{i}\xi_0 r_{\perp,\ell})\mathcal{B}_0\left[2\mathrm{i}|\boldsymbol{Q}_\parallel - \boldsymbol{g}_\parallel|\right]^{-1} \\ &\quad \times \left[\boldsymbol{K}_-^2 - |\boldsymbol{k}_{0,\parallel} + \boldsymbol{g}_\parallel|^2 - (k_{0,z} + \xi_0)^2\right]^{-1}\left[\boldsymbol{g}_\parallel^2 + \xi_0^2 + \lambda_\mathrm{eff}^2\right]^{-1} \end{aligned} \tag{6.19}$$

If $\boldsymbol{K}_-^2 > |\boldsymbol{k}_{0,\|} + \boldsymbol{g}_\||^2$ applies the function $\mathcal{L}_1$ reads

$$\begin{aligned}\mathcal{L}_1 &= \frac{-1}{2}\exp(-\mathrm{i}\xi_1 r_{\perp,\ell})\mathcal{B}_1\left[|\boldsymbol{Q}_\| - \boldsymbol{g}_\||^2 + (Q_z - \xi_1)^2\right]^{-1} \\ &\quad \times \left[\boldsymbol{K}_-^2 - |\boldsymbol{k}_{0,\|} + \boldsymbol{g}_\||^2\right]^{-1/2}\left[\boldsymbol{g}_\|^2 + \xi_1^2 + \lambda_{\mathrm{eff}}^2\right]^{-1} \end{aligned} \tag{6.20}$$

If however $\boldsymbol{K}_-^2 < |\boldsymbol{k}_{0,\|} + \boldsymbol{g}_\||^2$ is valid then the following equation applies

$$\begin{aligned}\mathcal{L}_1 &= \frac{-\mathrm{i}}{2}\exp(-\mathrm{i}\xi_1 r_{\perp,\ell})\mathcal{B}_1\left[|\boldsymbol{Q}_\| - \boldsymbol{g}_\||^2 + (Q_z - \xi_1)^2\right]^{-1} \\ &\quad \times \left[-\boldsymbol{K}_-^2 + |\boldsymbol{k}_{0,\|} + \boldsymbol{g}_\||^2\right]^{-1/2}\left[\boldsymbol{g}_\|^2 + \xi_1^2 + \lambda_{\mathrm{eff}}^2\right]^{-1} \end{aligned} \tag{6.21}$$

The function $\mathcal{L}_2$ has the form

$$\begin{aligned}\mathcal{L}_2 &= \frac{-\mathrm{i}}{2\sqrt{\boldsymbol{g}_\|^2 + \lambda_{\mathrm{eff}}^2}}\exp(-\mathrm{i}\xi_2 r_{\perp,\ell})\mathcal{B}_2\left[|\boldsymbol{Q}_\| - \boldsymbol{g}_\||^2 + (Q_z - \xi_2)^2\right]^{-1} \\ &\quad \times \left[\boldsymbol{K}_-^2 - |\boldsymbol{k}_{0,\|} + \boldsymbol{g}_\||^2 - (k_{0,z} + \xi_2)^2\right]^{-1} \end{aligned} \tag{6.22}$$

Expressions for the functions $\mathcal{B}_j,\ j = 1, 2, 3$ are obtained as follows

$$\mathcal{B}_j := (b_j - k_z)^{-1} + R(b_j + k_z)^{-1} - T(b_j - -\mathrm{i}\gamma)^{-1} \tag{6.23}$$

where $b_j := \boldsymbol{Q}_z - z_j + k_{2,z}$ and

$$\xi_0 = Q_z + \mathrm{i}\left[|\boldsymbol{Q}_\| - \boldsymbol{g}_\||^2 + \frac{1}{\lambda_{\mathrm{s}}^2}\right]^{1/2} \tag{6.24}$$

$$\xi_1 = \begin{cases} -k_{0,z} - \left[\boldsymbol{K}_-^2 - |\boldsymbol{k}_{0,\|} + \boldsymbol{g}_\||^2 - \mathrm{i}\eta\right]^{1/2}, & \text{if } \boldsymbol{K}_-^2 > |\boldsymbol{k}_{0,\|} + \boldsymbol{g}_\||^2 \\ -k_{0,z} + \left[\boldsymbol{K}_-^2 - |\boldsymbol{k}_{0,\|} + \boldsymbol{g}_\||^2 - \mathrm{i}\eta\right]^{1/2}, & \text{if } \boldsymbol{K}_-^2 < |\boldsymbol{k}_{0,\|} + \boldsymbol{g}_\||^2 \end{cases} \tag{6.25}$$

$$\xi_2 = \mathrm{i}(\boldsymbol{g}_\|^2 + \lambda_{\mathrm{eff}}^2)^{1/2} \tag{6.26}$$

Here $r_{\perp,\ell} < 0$ describes the interlayer distance that may depends on z, η is a small positive convergence factor, and λ_{s} is the TF screening length.

The amplitude T_{pe} can also be given in a closed analytical form as [68]

$$\begin{aligned} T_{\mathrm{pe}} = &\frac{-\mathrm{i}Z_{\mathrm{p}}}{\sqrt{2\pi V}}\frac{\delta(\boldsymbol{k}_\| - \boldsymbol{Q}_\| - \boldsymbol{k}_{2,\|})}{Q^2} \\ &\times\left[(Q_z + k_{2,z} - k_z)^{-1} + R(Q_z + k_{2,z} + k_z)^{-1} - T(Q_z + k_{2,z} - \mathrm{i}\gamma)^{-1}\right]\end{aligned} \tag{6.27}$$

In the present case of a parabolic dispersion of the conduction band electrons the integration of the $\boldsymbol{k}$ resolved cross section over the BZ is done analytically as

$$\sigma(\boldsymbol{k}_0, \Omega_2, E_2, \Omega_{\mathrm{p}}, E_{\mathrm{p}}) = \int \mathrm{d}^3\boldsymbol{k}\rho(\epsilon_k)f(\boldsymbol{k}, T)\sigma(\boldsymbol{k}_0, \boldsymbol{k}, \Omega_2, E_2, \Omega_{\mathrm{p}}, E_{\mathrm{p}}) \tag{6.28}$$

where $\rho(\epsilon_k)$ is the density of states at the temperature T and the energy ϵ_k which is associated with the wave vector k. $f(\boldsymbol{k}, T)$ is the Fermi–Dirac distribution. In the jellium model [$\rho = \rho_0 = V/4\pi^3$] and at $T = 0$ one finds thus that

$$\sigma(\boldsymbol{k}_0, \Omega_2, E_2, \Omega_\mathrm{p}, E_\mathrm{p}) = \rho_0 \int_{k \le k_\mathrm{F}} \mathrm{d}^3\boldsymbol{k}\,\sigma(\boldsymbol{k}_0, \boldsymbol{k}, \Omega_2, E_2, \Omega_\mathrm{p}, E_\mathrm{p}) \tag{6.29}$$

$$\sigma(\boldsymbol{k}_0, \boldsymbol{k}, \Omega_2, E_2, \Omega_\mathrm{p}, E_\mathrm{p}) = (2\pi)^4 \frac{k_p k_2}{v_0} |T_\mathrm{pe} + T_\mathrm{pc}|^2 (E_\mathrm{f} - E_\mathrm{i}) \tag{6.30}$$

Here $\Omega_\mathrm{p}, E_\mathrm{p}$ are respectively the solid scattering angle and the energy of the scattered projectile, as determined experimentally in the final channel. Some further numerical details are provided in [68]. If the incident projectile is an electron, one has to account for exchange effects within the two final-state electrons in the way described in Chapter 2, even if the sample is nonmagnetic. This means that the cross section in this case is a weighted average of the singlet and triplet cross sections.

The use of the TF model for the effective two-particle interaction enables us to obtain analytical results. However, as discussed in Appendix C.1 this approximation does not capture the frequency and the wave-vector dependent nature of the two-particle interaction in a polarizable medium. Below we will show that, away from the plasma frequency, the TF approximation leads indeed to physically reasonable results in a scattering experiment of the kind discussed in this work (Fig. 2.1). Hence, it makes sense to inspect within the TF model the dominant mechanisms leading to two interacting particles in the final channel. It should be stressed however that the calculational model we outline here relies on the validity of the first-order perturbative approximation (FBA) in the involved potentials, particularly in the electron–projectile interaction potential. As we demonstrated in Chapters 1 and 3, such an approximation provides reasonable results at high energies (compared to E_F). Due to screening, in solids the FBA is even more viable than in the case of atomic targets, however at low energies multiple scattering events may well have an influence on the pair emission intensity, as shown below.

6.2 Mechanisms of Correlated Electron Emission

6.2.1 Angular Pair Correlation

The local static form of the TF potential that we utilized for the coupling between the two particles and our experience from the previous chapters in analyzing the cross sections in terms of binary encounters suggest that we inspect the traces of possible (classical) kinematical scattering processes in the full quantum calculations.

Some of the possible classical collisions leading to the simultaneous emergence of two interacting particles in the final channel (the scattered projectile and the emitted electron) are depicted schematically in Fig. 6.1(a–d). The velocities of the vacuum particles are large with respect to the Fermi velocity, such that the model of the preceding section is applicable and for the classical analysis, we can assume the velocity of the conduction band electron to be negligible. On the scale of the interaction time this electron can then be viewed as being stationary. In the classical scattering processes shown in Fig. 6.1(a) the projectile is scattered into the vacuum following an inelastic binary encounter with the conduction band electron. The accelerated electron is elastically (totally) reflected from the crystal potential to escape into the vacuum with momentum $\boldsymbol{k}_2$. The scattered projectile emerges with a momentum $\boldsymbol{k}_\mathrm{p}$. In the quantum mechanical model of the preceding section, the interaction of the band electron with the crystal is facilitated by the initial-state binding. Classically, from the linear

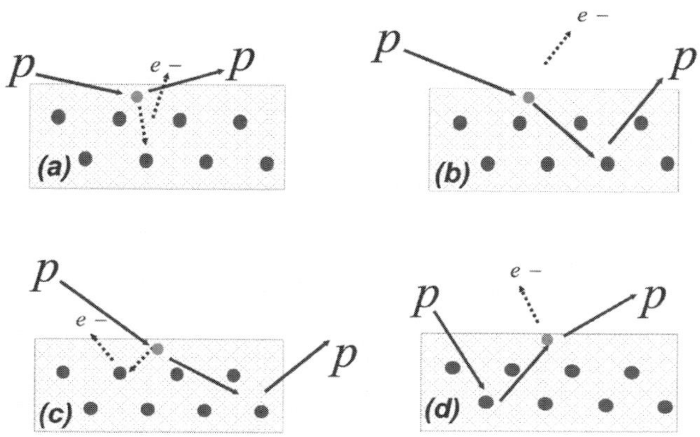

Figure 6.1: A schematic representation of some correlated two-particle scattering processes at surfaces. The wave vectors of the projectile (p) and the emitted electron (e^-) are indicated.

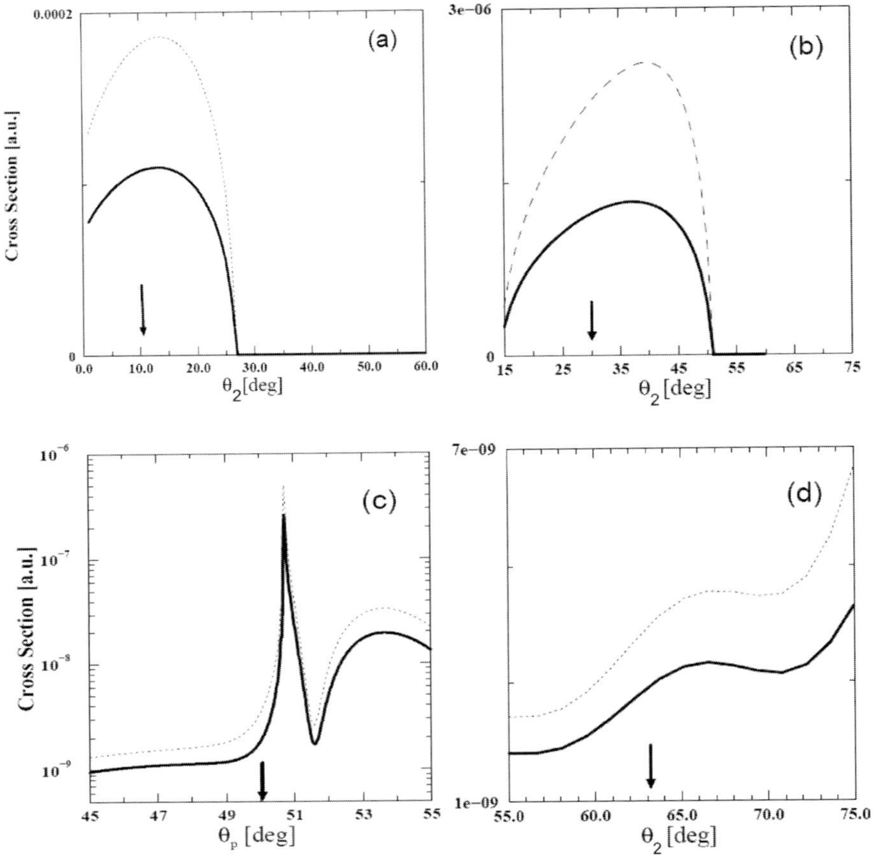

Figure 6.2: Coincident two-electron emission from an aluminum surface following the impact of 518 eV electrons. In (a), (b) and (d) the angular distributions of one electron, corresponding to the emitted electron in Fig. 6.1 are shown, whereas (c) shows the angular distribution of the scattered projectile electron (cf. Fig. 6.1). The emitted electrons emerge with an energy $E_2 = 60$ eV measured with respect to the vacuum level. The scattered electron possesses an energy of $E_{\mathrm{p}} = 453$ eV. The vectors $\boldsymbol{k}_0, \boldsymbol{k}_2$ and $\boldsymbol{k}_{\mathrm{p}}$ are chosen to lie in the same (x–z) plane (cf. Fig. 2.1). (a): The polar angles with respect to the surface normal are $\theta_0 = 100°$, $\theta_{\mathrm{p}} = 80°, \phi_0 = \phi_{\mathrm{p}} = \phi_2 = 0$, and θ_2 is variable. (b): As in (a) but $\theta_{\mathrm{p}} = 60°$. (c): As in (a) but $\theta_0 = 150°$, $\theta_2 = 40°, \phi_0 = \phi_{\mathrm{p}} = 0$, $\phi_2 = 180°$, and θ_{p} is variable. (d): $E_0 = 418$ eV, $E_{\mathrm{p}} = 302$ eV and $E_2 = 110$ eV, $\theta_0 = 178°$, $\theta_{\mathrm{p}} = 29°, \phi_0 = \phi_2 = 0$, $\phi_{\mathrm{p}} = 180°$, and θ_2 is variable. Calculations have been done including exchange (antisymmetrization) effects where the final state is described by the wave function (6.17) and the initial state is derived from the jellium model (6.15). The dashed curves stand for the results when including image–charge distortion in the calculations whereas in the calculations shown by the solid curves these distortions are neglected. The arrows in (a)–(d) indicate the positions of the peaks as predicted by the simple classical kinematics of the processes depicted in Fig. 6.1(a)–(d), respectively.

momentum conservation law we conclude that the mechanism shown in Fig. 6.1(a) results in a peak in the particles' angular distribution at $\theta_{\mathrm{p}} = \theta_0 - \arctan k_{\mathrm{p}}/k_2,\ \theta_2 = \pi - \theta_0 - \arctan k_{\mathrm{p}}/k_2, \phi_2 - \phi_{\mathrm{p}} = 0$. The polar angles are measured with respect to the surface normal (which defines the z axis). The x axis is along the surface and the x-z plane contains $\boldsymbol{k}_0$. The classical anticipations of the peak structures (marked by the arrows in Fig. 6.2(a)) are confirmed by the quantum mechanical calculations which are presented in Fig. 6.2(a). If we assume the conduction band electron to be stationary, a classical model anticipates a delta function peak in the angular distribution. In the quantum mechanical calculations however, the initial momentum distribution of the band electron results in a finite width of the peak observed in Fig. 6.2(a).

Figure 6.1(b) shows schematically a further classical collisional process: The conduction band electron is elevated to a vacuum state upon an inelastic encounter with the projectile. The scattered projectile is further scattered from the surface crystal potential. Classically this process is manifested in a delta-function peak structure in the angular distribution of the detected particles in the final channel at $\theta_{\mathrm{p}} = \pi - \theta_0 - \arctan k_2/k_{\mathrm{p}},\ \theta_2 = \theta_0 - \arctan k_{\mathrm{p}}/k_2,\ \phi_2 - \phi_{\mathrm{p}} = 0$. That this mechanism leaves traces in the calculated quantum mechanical cross section is confirmed by Fig. 6.2(b). Again the finite width of the momentum distribution of the conduction band electrons is reflected in a broadening of the classically predicted peaks.

Figure 6.2(c) illustrates the process in which both the electron and the projectile are totally reflected from the surface potential following a projectile–electron collision. Classically, the momentum conservation laws impose the following conditions under which the final state particles can appear $\theta_{\mathrm{p}} = \arctan k_2/k_{\mathrm{p}} + \pi - \theta_0, \theta_2 = \arctan k_{\mathrm{p}}/k_2 - \pi + \theta_0, \phi_2 - \phi_{\mathrm{p}} = \pm\pi$. The results of the quantum calculations shown in Fig. 6.2(c) endorse the influence of the classical scattering process depicted in Fig. 6.1(c). In this case we observe that the peak in the quantal results is quite sharp as compared to the peaks observed in Fig. 6.2(a), (b). The reason for this is that the displayed angular distribution corresponds to a much faster electron than in the case of Fig. 6.2(a), (b). Hence, the influence of the initial momentum distribution of the band electron becomes less prominent.

In Fig. 6.1(d) a further possible kinematical collision is displayed: the projectile is back reflected from the crystal potential and collides then with the electron. Both the electron and the projectile emerge then with momenta $\boldsymbol{k}_2$ and $\boldsymbol{k}_\mathrm{p}$ [6, 184–186]. Classically the process depicted in Fig. 6.1(d) results in a delta-function structure in the cross section at $\theta_2 = \arctan k_\mathrm{p}/k_2 + \pi - \theta_0$, $\theta_\mathrm{p} = \arctan k_2/k_\mathrm{p} + -\pi + \theta_0$, $\phi_2 = 0, \phi_\mathrm{p} = \pi$. These anticipations are also confirmed by the quantum mechanical calculations shown in Fig. 6.2(d). However, as a result of the low velocity of the electron whose angular distribution is scanned, we observe only a small shoulder at the classically predicted peak position.

Figure 6.2 illustrates that the inclusion of the image–charge potential in the calculations results in rather modified magnitudes of the two-particle coincidence intensity. The shapes of the angular distributions are, however, hardly affected.

6.2.2 Energy Pair Correlation

Experimentally, the angular distributions shown in Fig. 6.2 have not yet been measured with an accuracy that is sufficient to resolve the peaks predicted in Fig. 6.2. Extensive experimental and theoretical results exist for the energy correlation within the detected particle pair Refs. [44, 67, 68, 157, 185–188, 191]. Here we discuss the origin of the generic structure of the observed and calculated energy pair correlation.

An example is shown in Fig. 6.3 for Cu(001) and Ni(001) single crystals using an electron as a projectile. The theoretical results are obtained, as in Fig. 6.2, by calculating the scattering amplitudes T_pe and T_pc, antisymmetrizing and then summing over nonobserved quantum numbers. The crystal structure enters through the specific choice of the MT potential, however, as is clear from Fig. 6.3, the generic shape of the energy correlation within the emitted electron pair is the same for both targets and is not much dependent on the amount of the energy to be shared between the pair. Indeed more extensive theoretical and experimental results on other targets confirm this statement [157, 191].

The origin of the pronounced peak structure in the cross section when one electron is fast (at the wings of the distribution) can be traced back to the properties of the TF potential: Considering the scattering process (d) in Fig. 6.1 (which turned out to be dominant) one concludes, as shown in Chapter 1 (e.g., cf. Eq. (1.7)), that if the back-reflected projectile electron is fast (compared to v_F) the most probable outcome when it collides with the conduction band electron is one fast electron and one slow one. This is clear from Eq. (1.7) and from the momentum-space representation of the TF potential $\sim (q^2 + 1/\lambda_\mathrm{s}^2)^{-1}$, where q is the mo-

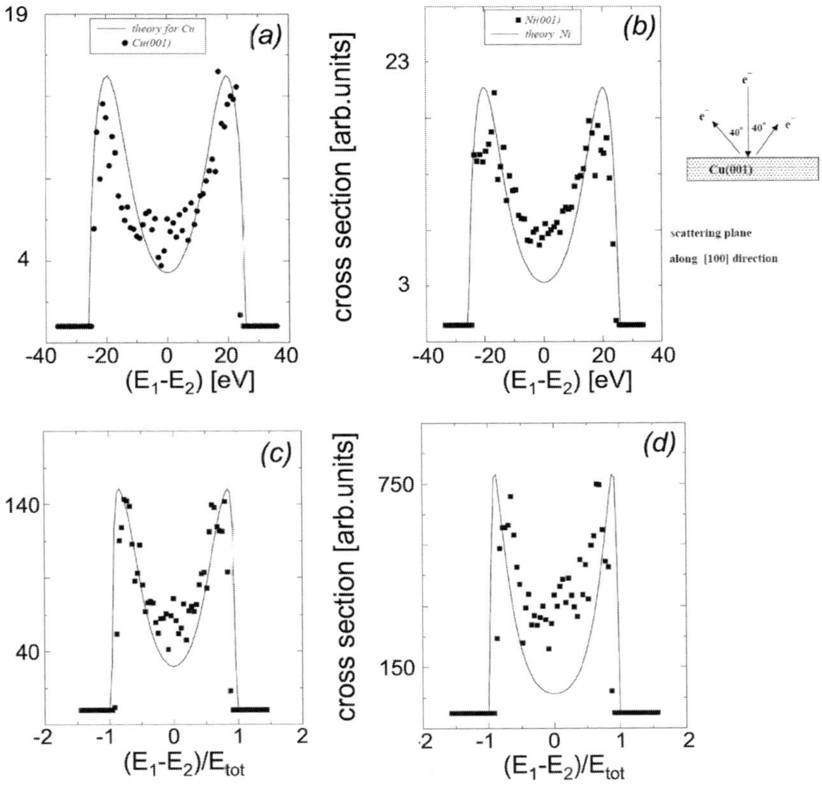

Figure 6.3: Simultaneous electron pair emission following the impact of unpolarized electrons in the geometry shown on the figure. The sharing of a fixed total energy $E_{\text{tot}} = E_1 + E_2$ between the two detected electron pair is shown. In (a) and (b) $E_{\text{tot}} = 27$ eV. The incident energy is $E_0 = 35$ eV. Experimental and theoretical results are shown for a Cu(001) crystal (a) and Ni(001) crystal (b). The incident beam is parallel to the surface normal. The two emitted electrons are detected at 40° to the left and to the right of the surface normal. The experimental data are determined on a relative scale only and have been normalized to the theoretical results at one point only. (c) and (d) are the same as in (a), however, the total energy of the pair is lowered respectively to $E_{\text{tot}} = 25$ eV and $E_{\text{tot}} = 23$ eV.

mentum transfer during the collision. From this form of the potential it is also clear that if the screening length is small (as compared to $1/q$) the TF potential becomes a flat function of q and so does the energy distribution.

The scattering properties of U_{eff} are reflected in the peak structure in Fig. 6.3 because the emission angles are chosen such that the condition $(\hat{\boldsymbol{k}}_1 \perp \hat{\boldsymbol{k}}_2)$ for the classical electron–electron encounter is nearly fulfilled. Away from this condition, i.e., changing the emission angles of the particles with respect to the surface normal (cf. Fig. 6.4) the form factor of the

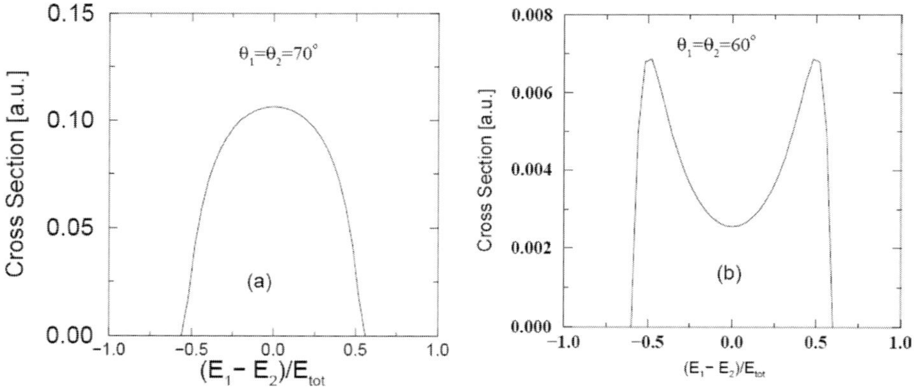

Figure 6.4: The same as in Fig. 6.3 however the emission angles θ_1 and θ_2 of the two electrons are varied as shown in the individual panels. The energies are $E_0 = 35$ eV and $E_{\text{tot}} = 30$ eV. The target is a Cu(001) single crystal.

potential becomes of less importance and we observe a flat energy-sharing distribution that drops to zero at the wings due to the vanishing phase space volume in this region (the cross section (6.30) is proportional to $\sqrt{E_1 E_2}$ and hence vanishes if $E_1 \to 0$ or $E_2 \to 0$). If $\tilde{U}_{\text{eff}}$ is smooth the shape of the energy-pair distribution is determined by the initial momentum distribution of the conduction band electron (cf. Eq. (1.7)). This feature is observed in Fig. 6.4: A single peak structure occurs in the energy-sharing distribution for the case of symmetric electron-emission angles with respect to the surface normal with $\theta_1 = \theta_2 = 70°$. When the emission angles are increased from $\theta_1 = \theta_2 = 40°$ (in Fig. 6.3) to $\theta_1 = \theta_2 = 70°$ we observe a shrinking of the cross section to a small energy region around $E_1 = E_2$ and eventually for $\theta_1 = \theta_2 = 70°$ one single peak at $E_1 = E_2$ is observed. This behavior can be traced back to the conservation law for the parallel components of the wave vectors and the range of wave vectors available in the initial-state momentum distribution: Considering the low-energy wings of the distributions, the surface-parallel wave vector of the electron which has the energy $E_1 \approx E_{\text{tot}}$ (or $E_2 \approx E_{\text{tot}}$) increases when increasing $\theta_1 = \theta_2$. The conservation law for the surface-parallel wave vector imposes, on the other hand, that the state of the conduction band electron must contain components with equally large surface-parallel wave vectors, otherwise the pair emission is prohibited. This means that the extent of the single peak in Fig. 6.4(a) is set by the width of the momentum space distribution of the band electron.

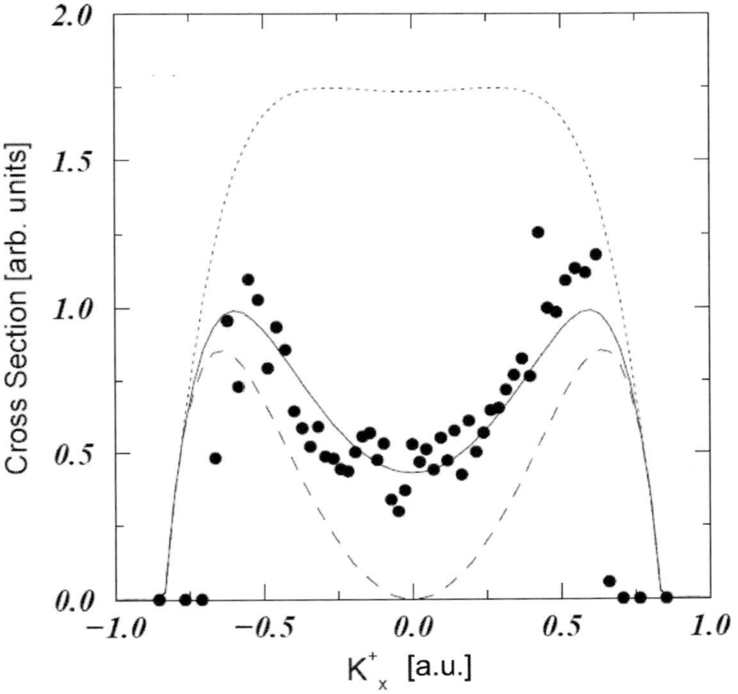

Figure 6.5: The same as in Fig. 6.3(a), however, $E_0 = 34$ eV and $E_{\rm tot} = 27$ eV. The singlet, (dotted curve) and the triplet (dashed curve) cross sections are shown along with their statistical average, i.e., spin-unresolved cross section (solid curve). The spin non-resolved experimental data (full dots) are on a relative scale and are normalized to theory at one point.

Another point worth mentioning here is the low-energy behavior, i.e., low $E_{\rm tot}$, of the energy pair distribution. In this context we recall our detailed discussion in Section 3.6, i.e., for small $E_{\rm tot}$ we expect that the spin-unresolved energy pair distribution will be dominated by singlet-pair emission.The structures arising due to the form of the potential are then of minor relevance. In general, exchange effects play a decisive role in determining the shape of the spin-unresolved cross section (6.30) in the low-energy regime.

6.2.3 Influence of Exchange on the Energy Pair Correlation

To illustrate how exchange effects govern the shape of the energy pair correlation we consider in Fig. 6.5 the singlet and the triplet contributions to the spin-unresolved cross section shown in Fig. 6.3(a). The cross sections are plotted as a function of the surface-parallel component of the center of mass wave vector $\boldsymbol{K}_x^+ = \boldsymbol{k}_{1x} + \boldsymbol{k}_{2x}$. Due to symmetry, both the singlet and the triplet contributions are symmetric with respect to $\boldsymbol{K}_x^+ = 0$ (cf. drawing in Fig. 6.3). The singlet contribution shows only a slight hint of the peak that stems from the form factor of the electron–electron scattering potential U_{eff}. The triplet scattering possesses a strict zero at $\boldsymbol{K}_x^+ = 0$. This is because, in the situation depicted in Fig. 6.3 when $\boldsymbol{K}_x^+ = 0$, an exchange of the two electrons is equivalent to a 180° rotation of the whole experimental setup around the surface normal. Since the experiment under study is invariant under such a rotation but the triplet scattering amplitude must change sign under an exchange of the two electrons we conclude that the triplet scattering vanishes at $\boldsymbol{K}_x^+ = 0$. We have already addressed this issue in a mathematically more strict way in Chapter 2. In fact Fig. 3.5 of Chapter 3 illustrates the similarity between the behavior of (the measured and the calculated) spin asymmetry in the atomic case and the case shown in Fig. 6.5. Meanwhile, there exists an impressive amount of spin-resolved energy-sharing distributions for magnetic surfaces confirming that the shape of the spin asymmetry in the case of the pair emission from spin-polarized surfaces resembles to some extent the atomic case shown in Fig. 2.3 [156, 157, 203]. However, since the spin-polarization of the electronic states of the sample changes (or vanishes) with the binding energy (cf. for instance Fig. 2.2) the sign of the spin asymmetry depends strongly on the initial energy of the band electron.

The situation presented in Fig. 6.5 demonstrates that, regardless of the fact that the calculations and the experiment of Fig. 6.5 are spin-unresolved, at $E_1 = E_2$ only singlet electron pairs are emitted. This is valid if $\theta_1 = \theta_2$ and provided that the crystal structure is invariant under 180° rotation around $\hat{\boldsymbol{z}}$. Moreover, Fig. 6.5 evidences that the pronounced peaks observed in Figs. 6.3–6.5 at the wings of the distributions are in fact due to three factors:

1. The behavior of the electron–electron scattering potential $\tilde{U}_{\text{eff}}$ plays an important role, as detailed above.

2. Exchange effects dictate zero points in the triplet cross sections at $E_1 = E_2$ which results in depressions at $E_1 = E_2$ in the spin-averaged cross sections.

3. Due to kinematical reasons the cross section (6.30) vanishes if $E_1 \to 0$ or $E_2 \to 0$. This is decisively different from the case of a neutral atomic target. For an atomic target in

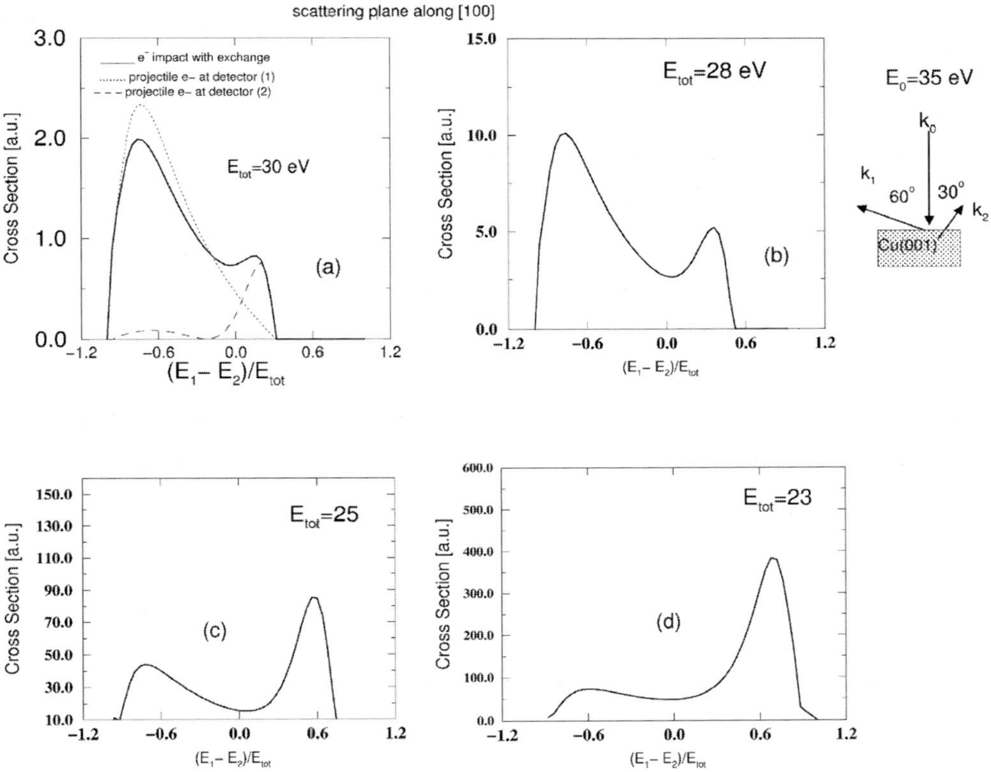

Figure 6.6: The spin-averaged energy-sharing distributions within an electron pair emitted from a Cu(001) single crystal following the impact of 35 eV electrons. The variable total kinetic energy of the pair $E_{\rm tot}$ is shown on the respective figures. As shown schematically, one of the detectors (labelled 2) is fixed at an angle of 30° to the right of the surface normal and the other detector (detector 1) is positioned at 60° to the left of $\boldsymbol{k}_0$. The solid curves in (a)–(d) are the spin-averaged theoretical cross sections. In (a) the calculations represented by the dotted and by the dashed curves do not account for exchange effects. The dotted and the dashed curves represent the theoretical predictions if the projectile electron is detected with detector 1 or 2, respectively.

the region $E_1 \to 0$ or $E_2 \to 0$ the single-particle density of states of the emitted electron diverges as $1/\sqrt{E_1}$ which, in combination with the $\sqrt{E_1 E_2}$ kinematical dependence of Eq. (6.30), leads to a finite cross section when $E_1 \to 0$ or $E_2 \to 0$.

Away from the highly symmetric situation $\theta_1 = \theta_2$ of Fig. 6.3 the exchange effect may still have a strong influence on the correlated pair emission. This is demonstrated in Fig. 6.6 where the emission angles are unequal ($\theta_1 = 30°$ and $\theta_2 = 60°$). Now let us neglect the influence of exchange[4]: If the scattered projectile electron is detected to the right of the incoming beam (i.e., with an emission angle of $30°$) we obtain the theoretical results shown by the dashed curve in Fig. 6.6(a). On the other hand, if the scattered electron emerges in the final channel to the left of the surface normal the cross section has the behavior described by the dotted curve in Fig. 6.6(a). The peak structures in the solid and dashed curves in Fig. 6.6(a) have their origins in the properties of $\tilde{U}_{\text{eff}}$ as discussed above. When including exchange effects in the calculation (results are shown by the solid curve in in Fig. 6.6 (a)) we conclude that the two maxima in the spin-unresolved cross section stem from the corresponding peaks in the direct and in the exchange scattering amplitudes.

Figure 6.6(a) also demonstrates one anticipation that we expressed by Eq. (2.39) in Chapter 2: when the two electrons have substantially different wave vectors exchange effects play a subsidiary role, meaning that at higher energies the spin-unresolved cross section merges with the cross section built out of the direct scattering amplitude. Fig. 6.6(a) shows vestiges of this trend even at moderate energies.

Figure 6.6(b)–(d) shows the influence of varying the initial crystal momentum distribution. This is done by fixing the impact energy E_0 and lowering the total kinetic energy of the pair E_{tot}. In this way, due to the energy conservation law, the sample electron originates from deeper levels in the conduction band. In Fig. 6.6(a)-(d) we observe a strong dependence on E_{tot} which endorses the decisive role of the initial-momentum distribution. This variation of the cross section with E_{tot} cannot be obtained from a classical argument and becomes less pronounced with increasing incident energy, i.e., when E_0 is much larger than the conduction band width.

6.2.4 Pair Diffraction

On the basis of a general analysis we pointed out in Section 6.1.3 the possibility of the diffraction of an interacting two-particle compound. In a number of experiments this prediction has been confirmed [67, 204]. Here we illustrate some typical examples (Fig. 6.7) of how the pair diffraction shows up in the energy pair distribution. In the geometry of the experiment sketched in the inset of Fig. 6.7. All vacuum wave vectors, i.e. $\boldsymbol{k}_0, \boldsymbol{k}_1$, and $\boldsymbol{k}_2$ lie in the same

[4]In this case the cross section in the geometry of Fig. 6.6 is equivalent to the cross sections for the coincident electron emission upon positron impact. This is because the amplitude T_{pe} makes almost no contribution to the cross section (as checked numerically) and $|T_{\text{pc}}|^2$ is proportional to Z_{p}^2.

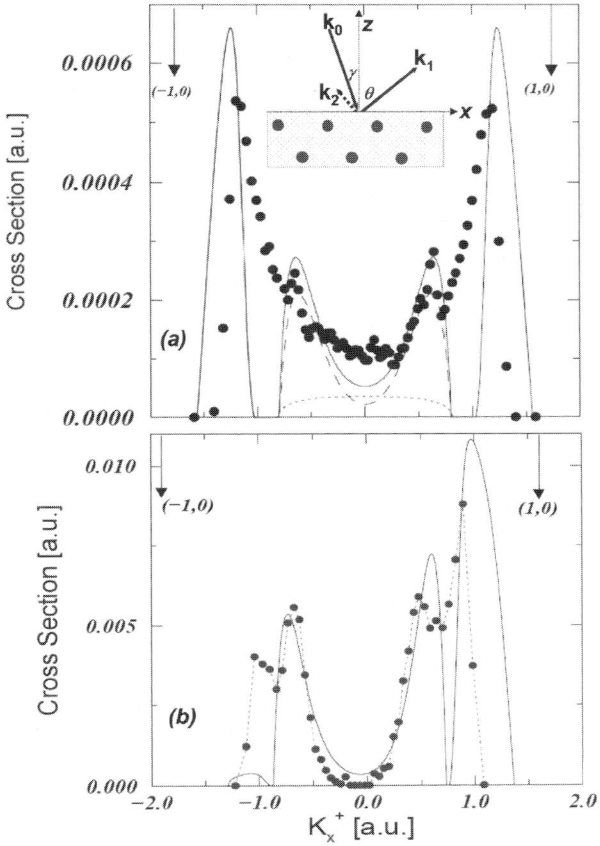

Figure 6.7: The spin-unresolved energy-sharing distribution within an electron pair emitted upon the impact of an electron beam on a single crystal in the geometry sketched in the inset. (a): $E_0 = 85$ eV and the total kinetic pair energy is $E_{\text{tot}} = 79$ eV. Experiments and calculations have been conducted for a Cu(001) crystal with normal incidence of the incoming electron beam, i.e., for $\gamma = 0$. The emission angles of both electrons is $\theta = 40°$. The figure also shows the separate (incoherent) contributions of the electron–electron scattering amplitudes $|T_{\text{pe}}|$ (dotted curve in (a)), as given by Eq. (6.13), and the electron–crystal scattering amplitude $|T_{\text{pc}}|$ (dashed curve in (a)) that describes the pair's scattering from the lattice potential [cf. Eq. (6.12)]. The results of the coherent sum $|\mathcal{T}| = |T_{\text{pe}} + T_{\text{pc}}|$ (solid curve) are also depicted. The theory does not account for the finite experimental resolution. The theoretical (-1,0) and (1,0) diffracted peaks have been scaled down by a factor of 2. The experimental data [67] (full dots) are on a relative scale and have been normalized to the theoretical results at one point. (b): The experimental results (full dots) for a Fe(110) (BCC) crystal. The incident energy is $E_0 = 50$ eV and $E_{\text{tot}} = 44$ eV. The incoming electron beam is tilted at an angle of $\gamma = 5°$ with respect to the surface normal. The emission angles are $\theta = 50°$. The spin-averaged theoretical results (solid curve) are obtained by evaluating $|\mathcal{T}| = |T_{\text{pe}} + T_{\text{pc}}|$ (Eq. (6.1)).

x-z plane (normal to the surface). Therefore, the pair center of mass wave vector $\boldsymbol{K}_{\parallel}^{+}$ has only one nonvanishing component, namely along the x axis. The energy sharing is then studied as a function of this component K_x^+. Considering the two-particle system as a composite particle with a total mass m and a reduced mass m_μ the total kinetic energy of the emitted particle reads then $E_{\text{tot}} = {K^+}^2/(2m) + {K^-}^2/(2m_\mu)$ ($m_\mu = 0.5$ for two electrons). The magnitude of K_x^+ is restricted to the interval $0 \leq K_x^+ \leq \sqrt{4E_{\text{tot}}}$. From Fig. 6.7 we infer then the following condition

$$-\sin\theta\,\sqrt{2E_{\text{tot}}} \leq K_x^+ \leq \sin\theta\,\sqrt{2E_{\text{tot}}}. \tag{6.31}$$

Figure 6.7(a) shows the spin-unresolved K_x^+ distribution of the coincident two-electron emission cross section (6.30) from a Cu(001) single crystal. The Bloch wave vector $\boldsymbol{k}_{\parallel}$ of the initially bound electron is unknown and should be summed over. For the sake of interpretation however, let us assume $\boldsymbol{k}_{\parallel} = 0$. Then from the structure of the transition amplitude T_{pc} we conclude that the first diffraction maxima are positioned at the values of K_x^+ which are indicated on Fig. 6.7 by arrows marked $(-1,0)$ $(1,0)$ maxima. Both the theoretical and the experimental results evidence the onset of the $(1,0)$, and $(-1,0)$ diffraction peaks. These peaks are however not fully developed since the cut-off condition (6.31) restricts the allowed K_x^+ values. The specular beam $(0,0)$ results in the double-peak structure located between $-1 \lesssim K_x^+ \lesssim 1$ in Fig. 6.7. This statement follows from a separate analysis of the scattering amplitudes for the direct projectile–electron encounter $|T_{\text{pe}}|$ and the amplitude $|T_{\text{pc}}|$ that describes the projectile crystal-lattice scattering accompanied by a projectile–electron scattering as well as the coherent sum amplitude (6.1). As is clear from Fig. 6.7(a) the contribution to the pair emission cross section from T_{pc} (Eq. (6.12)) is dominant. The amplitude T_{pe} (Eq. (6.13)), which does not involve scattering from the crystal potential contributes only in the region of the specular beam.

In Section 6.1.3 we also concluded that the width of the $(0,0)$ diffraction structure is governed by the x component of the initial total wave vector of the pair. Since the incident wave vector $\boldsymbol{k}_0$ is determined by experiment the width of the pair initial momentum distribution stems from the wave-vector spread of the bound electron. The maximal value of the wave-vector components present in the initial state is given the Fermi wave vector k_{F}. Consequently we conclude that the diffraction peaks extend in K_x^+ space over a region of $\pm k_{\text{F}}$. We recall however, that the cut-off condition (6.31) may restrict the allowed widths of the diffraction peaks.

Figure 6.7(b) illustrates yet another situation for a Fe(110) single crystal: Here the lattice structure of the sample and the scattering plane do not possess a common symmetry axis. In this case we have a situation similar (from a symmetry point of view) to the case we encountered in Fig. 6.6. In contrast to Fig. 6.7(a) the cross sections depicted in Fig. 6.7(b) are not symmetric with respect to $K_x^+ = 0$. In addition, contrary to Fig. 6.7(a) where $k_{0,x} = 0$ in Fig. 6.7(b) $k_{0,x} = 0.17$ a.u. and hence the initial distribution of the x component of the pair total wave vector $K_{0,x}$ is modified in a preferential direction. Hence the positions of the $(-1,0)$ and $(1,0)$ diffraction peaks in Fig. 6.7(b) are not symmetric with respect to $K_x^+ = 0$. It should be noted however that the asymmetry of the energy sharing distributions is inherent to the behavior of exchange effects, as we inferred from the analysis of Fig. 6.6.

6.3 Role of the Dynamical Collective Nature of the Two-particle Interaction

For the preceding calculations we employed for the effective two-particle interaction $\tilde{U}_{\text{eff}} = U/\epsilon$ a statically and locally screened Coulomb interaction. As discussed in Appendix C.1, this amounts to the use of an approximate dielectric function that has the form $\epsilon(q,\omega = 0) = 1 + \frac{\lambda_s^2}{q^2}$, where ω is the frequency. However, as we demonstrated in Chapter 5 the retarded nature of $\tilde{U}_{\text{eff}}(q,\omega)$ may well have a profound influence on the measured two-particle correlation functions.

In this section we study therefore the role of the frequency and the wave-vector dependent nature of $\tilde{U}_{\text{eff}}(q,\omega)$ in determining the behavior of the cross sections for the correlated two-particle emission. The findings are contrasted with the results of the TF approach. For this purpose we utilize the modified dynamical Lindhard–Mermin dielectric functions [205, 206] (LM) which has the following explicit form

$$\begin{aligned}
\epsilon(q,\omega + \mathrm{i}\Gamma) &= 1 + N/D \\
N &= (\omega + \mathrm{i}\Gamma)\left[\epsilon_{\mathrm{L}}(q,\omega + \mathrm{i}\Gamma) - 1 + 4\pi\chi_{\text{core}}\right] \\
D &= \omega\left\{1 - G(q)[\epsilon_{\mathrm{L}}(q,\omega + \mathrm{i}\Gamma) - 1]\right\} + \mathrm{i}\Gamma\left\{1 - G(q)[\epsilon_{\mathrm{L}}(q,0) - 1]\right\} \\
&\quad \times \frac{\epsilon_{\mathrm{L}}(q,\omega + \mathrm{i}\Gamma) - 1 + 4\pi\chi_{\text{core}}}{\epsilon_{\mathrm{L}}(q,0) - 1 + 4\pi\chi_{\text{core}}}
\end{aligned} \tag{6.32}$$

Here $\epsilon_{\mathrm{L}}(q,\omega)$ is the Lindhard dielectric function [207] (cf. Appendix C).

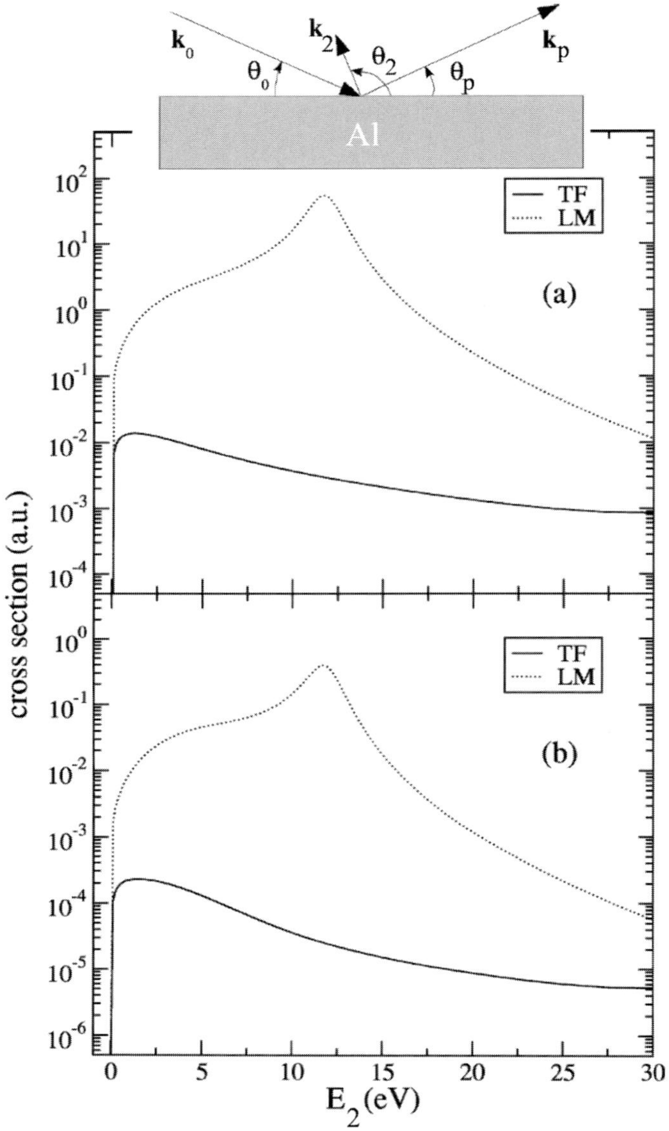

Figure 6.8: The coincident electron emission from the (001) fcc face of aluminum following the impact of a proton. The scattering geometry is illustrated schematically in (a). The energy distribution is plotted for an electron that emerges normal to the surface. The MT potential is derived from DFT-LDA calculations. In (a) we consider the case of a specular reflection of the proton, i.e., $\theta_0 = \theta_p = 15^\circ$; whereas in (b) the situation of a non-specular reflection mode, namely $\theta_0 = 75^\circ, \theta_p = 15^\circ$ is considered. The energies of the proton in the initial and in the final channel are respectively $E_0 = 100$ keV and $E_p = E_0 - E_2 - W$.

Below we discuss numerical results for aluminum. For Al the parameters entering Eq. (6.32) are [205]

$$\Gamma = (0.53 + 30.9Z^2\,\mathrm{eV})\Theta(0.067 - Z^2) + (2.6 + 0.2\,\mathrm{eV})\Theta(Z^2 - 0.067),$$
$$G = 2.5Z^2 - \mathrm{i}[2.12Z^2\Theta(0.067 - Z^2) + 0.142\Theta(Z^2 - 0.067)],$$
$$4\pi\chi_{\mathrm{core}} = 0.05, \quad Z = q/2k_{\mathrm{F}} \tag{6.33}$$

$\Theta(x)$ denotes the step function.

Figure 6.8(a) shows the cross section as a function of the energy E_2 of one electron emitted from an aluminum (001) single crystal surface upon the impact of a proton as a projectile (a schematic drawing is shown in Fig. 6.8). The electrons are emitted normal to the surface whereas the protons are specularly reflected at the sample surface. To understand qualitatively the shape of the energy distribution displayed in Fig. 6.8 we recall Eq. (1.7) which states that the cross section for the two-particle emission is, in a rough approximation, proportional to the interaction potential between the particles $\tilde{U}_{\mathrm{eff}}$. As is clear from Eqs. (6.32) and (C.7) the TF and LM models treat $\tilde{U}_{\mathrm{eff}}$ profoundly differently with the following consequences for the cross sections:

1. The cross sections evaluated using the LM dielectric function are two orders of magnitude larger than those obtained when the TF theory is utilized.

2. The electron energy distribution calculated using the LM model reveals a maximum at $E_2 \approx \omega_{\mathrm{pl}} - W$, where ω_{pl} is the plasma frequency. The energetic position and the width of this peak are governed by, respectively, the energetic position of the plasmon pole and the plasmon life time (cf. Appendix C.3). Obviously, the static local FT model of screening is not capable of reproducing this feature.

It is worthwhile to note that the plasmon wave vector is small (< 1 a.u.) on the scale of the wave vector of the incoming projectile. To excite this collective electronic mode the energy transfer to the electronic charge cloud should thus be small during the proton scattering. This can be achieved through the choice of an appropriate scattering geometry, as demonstrated in Fig. 6.8.

We note however, that the momentum transfer to the whole sample could however be substantially large (as is the case in Fig. 6.8(a)). This momentum is transferred however mainly to the surface-parallel atomic layers before and/or after the proton–electron interaction.

Figure 6.8(b) shows the geometry of a nonspecular reflection of the incident proton. In contrast to the case of specular reflection, the momentum is transferred not only to the surface-parallel atomic layers but also to the surface-perpendicular planes. Also in this case a small amount of momentum can be absorbed by the electronic charge cloud leading to a plasmon-assisted two-particle scattering. For this reason the electron-energy distributions shown by Fig. 6.8(a) and (b), have similar shapes both in the specular and in the nonspecular reflection of the proton from the sample surface. In Fig. 6.8(b) we observe a substantially decreased cross section as compared to Fig. 6.8(a). This is a manifestation of the unfavored scattering through large angles. More examples that demonstrate in particular the influence of the dynamical screening on the angular correlation function can be found in [47].

Apart from the plasmon-related features the cross section calculated with the LM and TF model of screening differ mainly in magnitude. The qualitative trend of the shape, away from the plasmon pole, is in both cases rather similar. Hence the TF model may well serve as a useful tool for a semi-quantitative description of the shape of the cross sections which is valuable in view of the fact that the experiments available till now do not determine the absolute value of the cross sections. The conclusion concerning the usefulness of the TF model may seem surprising in view of its known shortcomings in describing correctly charge-density modulation phenomena, such as Friedel oscillations. The difference in our case is that the scattering geometry (i.e., the energies and emission angles of the chosen particles) may be set up such that distant collisions are most relevant. In this case the TF is a reasonable approximation. On the other hand from this scenario it is clear that for head-on (close) collisions the long-wave length TF approximation is no longer justifiable.

6.4 Quantitative Description of Pair Emission from Surfaces

The theory we employed in the previous sections and in fact most theories devised so far for the treatment of the correlated two-particle emission from surfaces give the electron–electron interaction U_{eff} to a first order, as far as the scattering between the escaping particles is concerned. Such an undertaking might be inadequate, in particular when the particles emerge with comparable momenta. In addition, at higher energies ($> E_{\text{F}}$) screening of U_{eff} by the surrounding medium (mean field) might be less effective, thus necessitating the inclusion of higher-order scattering processes involving U_{eff}. A general conceptual and numerical approach to circumvent this problem is still lacking. A rather simple pragmatic way to explore the effect of strong two-particle correlation is to incorporate U_{eff} as a dynamical modification of the effective single particle potentials V_j, $j = 1,2$ experienced by the two escaping particles, i.e. V_j, $j = 1,2$ are then strictly speaking many-particle potentials.

6.4.1 Treating Strong Two-particle Correlations

The influence of exchange (between the two particles) is taken exactly into account if one uses antisymmetrized initial $(|\varphi_{\epsilon_i}(\boldsymbol{r}_2)\psi_{\boldsymbol{k}_0}(\boldsymbol{r}_1)\rangle\rangle)$ or final-state two-particle wave functions $(|\Psi_{\boldsymbol{k}_1,\boldsymbol{k}_2}(\boldsymbol{r}_1,\boldsymbol{r}_2)\rangle\rangle)$. Hence, we concentrate here on the Coulomb correlation.

The scattering from the crystal potential and all other electrons, as well as inelastic scattering processes, can be treated on the single particle level. Furthermore, spin-related effects can also be incorporated in the theory on the single particle level. This is achieved by deriving the wave function of the particle j by solving for the Dirac equation involving the optical potential V_j. Within the local-density approximation to the density functional theory the effective single particle potential $V_j(\boldsymbol{r}_j)$ has the (local) form

$$V_j(\boldsymbol{r}_j) = V_{\text{H}}(\boldsymbol{r}_j) + v_{\text{ex}}(\boldsymbol{r}_j) + V_{\text{O}}(\boldsymbol{r}_j) \tag{6.34}$$

where V_{H} is the Hartree potential and $v_{\text{ex}}(\boldsymbol{r}_j)$ is the exchange and correlation potential. $V_{\text{H}}(\boldsymbol{r}_j) + v_{\text{ex}}(\boldsymbol{r}_j)$ determines the ground state properties of the sample and can be calculated by means of a self-consistent band structure method, such as the linear muffin-tin orbital (LMTO) method. The optical potential V_{O} accounts (phenomenologically) for inelastic scattering processes and is specified in some details in the next chapter.

The lattice periodic potential V_j can be expressed in a MT form and the Korringa–Kohn–Rostoker (KKR) method [42, 151, 208–210] can be utilized for calculation of the single-particle relativistic scattering state $|\psi_{E_j,\boldsymbol{k}_{j\|},m_{s_j}}\rangle$, commonly known as the low-energy electron diffraction, LEED state. Here m_{s_j} specifies the electron spin state. We note that even at low energies the crystal potential (6.34) may contain a spin-dependent component (e.g., spin–orbit) coupling which entails the use of spin-dependent LEED states to capture correctly the effects of the spin-split band structure of the sample.

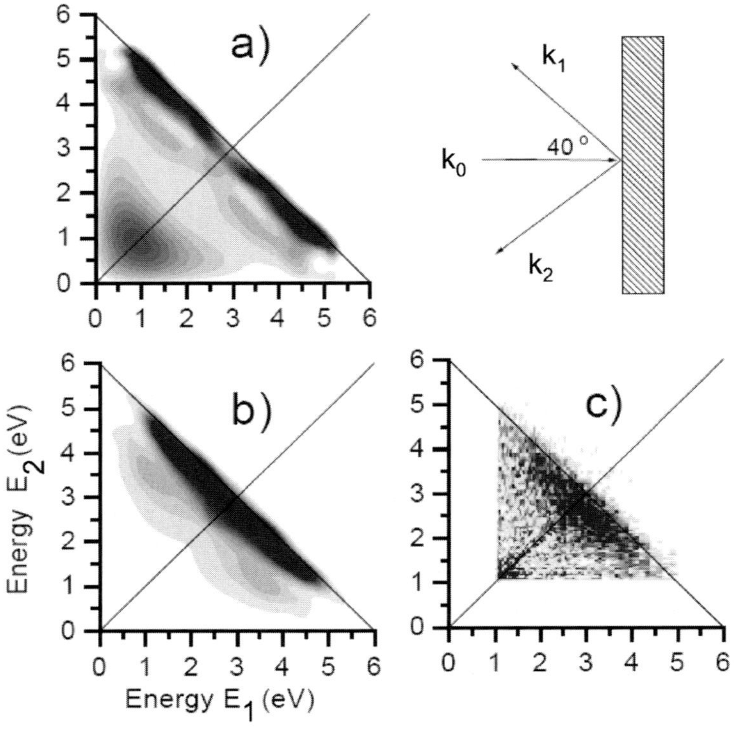

Figure 6.9: The two-particle energy distribution of two correlated spin-unresolved electrons emitted simultaneously upon the impact of a 10 eV spin-unpolarized electron onto the W(001) surface. $\boldsymbol{k}_0$ is aligned along the surface normal which defines the z axis. The x axis is chosen along the direction [100] of the sample. All momenta of the electrons $\boldsymbol{k}_0, \boldsymbol{k}_1, \boldsymbol{k}_2$ are chosen to be in one plane, the x-z plane. As shown schematically, the polar angles are fixed at $\theta_1 = \theta_2 = 40°$ and the azimuthal angles are $\phi_1 = 0$ and $\phi_2 = 180°$. Density plots show the results of the LKKR calculations when $U_{\text{eff}} \equiv 0$ in Eq. (6.35) (panel (a) whereas in panel (b) the full potential, as given by Eq. (6.35) is taken into account (under the approximation (6.38)). Experimental data are shown in panel (c). The energy regimes for which $\epsilon_i = E_{\text{F}}$ and $E_1 = E_2$ are indicated respectively by the solid counter-diagonal and the diagonal lines. See also color figure on page 150.

As done above and in most studies the two-particle state $|\Psi_{\boldsymbol{k}_1,\boldsymbol{k}_2}(\boldsymbol{r}_1,\boldsymbol{r}_2)\rangle$ at the total energy $E_1 + E_2$ is approximated by a direct antisymmetrized product of two LEED states, $|\psi_{\boldsymbol{k}_1}(\boldsymbol{r}_1)\rangle$ and $|\psi_{\boldsymbol{k}_2}(\boldsymbol{r}_2)\rangle$ [42, 190]. The electron–electron interaction U_{eff} is treated perturbationally which leads to a Golden-rule expression (cf., e.g., Eq. (6.2)) for the transition amplitude. In principle however, $|\Psi_{\boldsymbol{k}_1,\boldsymbol{k}_2}(\boldsymbol{r}_1,\boldsymbol{r}_2)\rangle$ should follow from a solution with appropriate boundary condition of a two-particle Dirac equation involving the total potential [211,212]

$$V_{\text{tot}} = V_1 + V_2 + U_{\text{eff}} \tag{6.35}$$

Subsequently, the S matrix elements (4.26) should be evaluated. Performing such as a procedure exactly is highly demanding and has not yet been done.

On the other hand, as we demonstrated in Chapter 3 the method of effective charges may well serve as a useful tool to incorporate higher-order interactions. In Chapter 3 we introduced effective charges in order to capture the influence of three and higher-order coupling terms while all two-particle multiple scattering processes were treated exactly. Here we may employ the same procedure to incorporate the influence of the two-particle interaction U_{eff} on the single-particle states $|\psi_{\boldsymbol{k}_1}(\boldsymbol{r}_1)\rangle$ and $|\psi_{\boldsymbol{k}_2}(\boldsymbol{r}_2)\rangle$ [212,213].

To this end we utilize for U_{eff} the TF form $U(\boldsymbol{r}_1, \boldsymbol{r}_2) = [\exp(-r_{12}/\lambda_{\text{s}})]/r_{12}$. The value of the screening length λ_{s} follows from its functional form (at $T = 0$) for the homogeneous electron gas, where the only undetermined parameter is the total density of states $\text{dos}(E_{\text{F}})$ at the Fermi energy (cf. Appendix C.1). For (real) inhomogeneous solids we use however the value for $\text{dos}(E_{\text{F}})$ as derived from the self-consistent electronic structure calculation for the crystal under study.

For a further advance the one-electron potentials V_1 and V_2 are expanded as lattice sums over potentials w_1 and w_2 that act within MT spheres located around the lattice sites. Within each sphere the total potential is cast as $W_{\text{tot}} = w_1 + w_2 + U_{\text{eff}}$. In the spirit of the effective charge method the potential W_{tot} is reformulated as follows

$$W_{\text{tot}} = w_1 + \frac{Z_1}{r_1} + w_2 + \frac{Z_2}{r_2} = \bar{w}_1 + \bar{w}_2 \tag{6.36}$$

$$\begin{aligned} Z_j &= a_j^{-1} \exp\left(-\frac{a_j}{2\lambda_{\text{s}}} r_j\right), \quad j = 1, 2, \\ a_j &= 2r_{12}/r_j \end{aligned} \tag{6.37}$$

This exact reformulation does not lead to any simplification of the calculations. An advance in this respect is achieved by recasting the dynamical screening functions Z_j in an approximate way

$$\begin{aligned} &\bar{Z}_j = \bar{a}_j^{-1} \exp\left(-\frac{\bar{a}_j}{2\lambda_{\text{s}}} r_j\right) \\ &\bar{a}_1 = 2v_{12}/v_1, \quad \bar{a}_2 = 2v_{12}/v_2, \quad \boldsymbol{v}_{12} = \boldsymbol{v}_1 - \boldsymbol{v}_2 \end{aligned} \tag{6.38}$$

where $\boldsymbol{v}_1$ and $\mathbf{v}_2$ are the velocities of the two vacuum electrons.

The approximation (6.38) for Z_j means that we assumed $\boldsymbol{r}_1 \propto \boldsymbol{v}_1$ and $\boldsymbol{r}_2 \propto \boldsymbol{v}_2$. Strictly speaking this approximation is justified if the potentials in the MT sphere are smooth and weak (in between the MT spheres the potential vanishes). In this case one may assume classically that the positions of the particles are proportional to their velocities.

For the calculations shown in Fig. 6.9 and 6.10 the velocities are chosen such that $\boldsymbol{v}_j = \hbar \boldsymbol{k}'_{j0}/m,\ j = 1, 2$, where $\boldsymbol{k}'_{j0}$ is the wave vector of the 00 LEED beam inside the crystal associated with the wave vector $\boldsymbol{k}_{j0}$ outside the crystal upon a refraction at the surface potential step.

From a physical point of view Eq. (6.36) means that the electronic correlation modifies the single particle potentials $w_j,\ j = 1, 2$ in the way dictated by Z_j/r_j. The term Z_j/r_j introduces in the single particle potential a dependence on the wave vectors of the two particles as well as on the interparticle wave vector. The way in which the features akin to interparticle correlation are reflected in the single particle properties is inferred from the behavior of the functions $\bar{Z}_j$, as given by Eq. (6.37). When the two excited particles approach each others in momentum space, i.e., when $v_{12} \rightarrow 0$ the potentials $\bar{w}_j = w_j + Z_j/r_j$ become repulsive. In contrast, if the two electrons possess substantially different velocities $v_i \gg v_j,\ i \neq j \in [1, 2]$ the screening strengths $\bar{Z}_1$ and $\bar{Z}_2$ tend to zero and we recover, as expected, the independent particle description. Another limit which should be (and is) ensured is that when one particle is strongly coupled to the ionic sites then its motion becomes governed by the corresponding ionic site potential w_j ($\lim_{v_j \rightarrow 0} \bar{w}_j \rightarrow w_j$).

6.4.2 Relativistic Layer KKR Method

With the aid of Eqs. (6.38) the potentials $\bar{w}_j$ turn local (but change with energy) and depend on one spatial coordinate ($\boldsymbol{r}_j$) only. Hence we can now use these potentials for the relativistic layer-KKR numerical calculations of the (modified) time-reversed LEED states out of which we construct the two-particle LEED state as an antisymmetrized direct product. The screening length λ_s is derived from a self-consistent LMTO calculation [193]. For tungsten, which is studied here, we obtain $\lambda_\mathrm{s} = 0.48$ Å.

The KKR method as such is well documented in the literature (cf. for example [197, 202, 214–218] and references therein). In brief the semi-infinite solid is decomposed in surface-parallel planes with the vacuum half space being separated from the solid by an interface

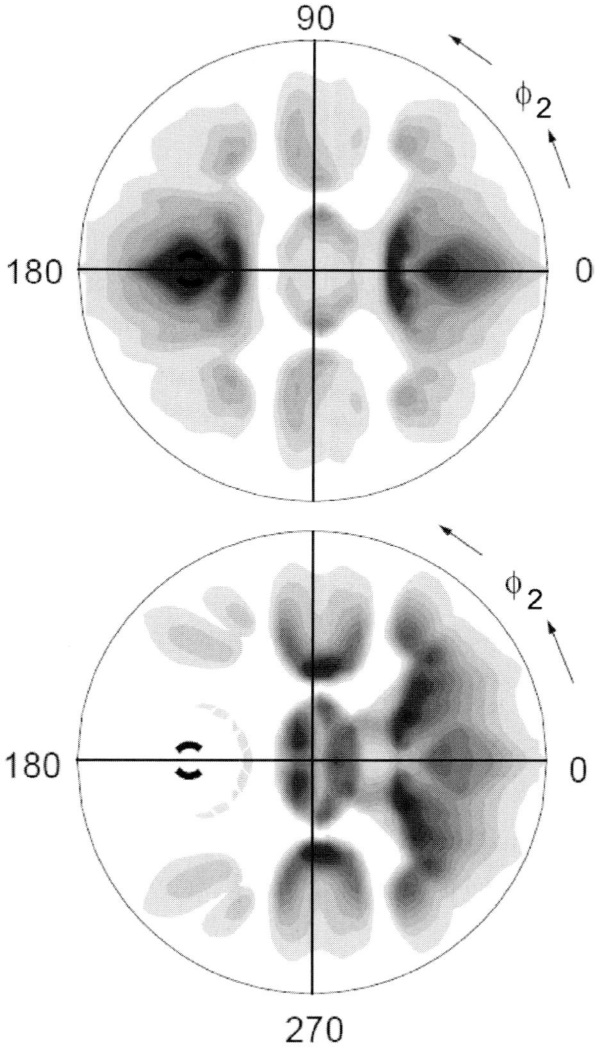

Figure 6.10: The results of the LKKR calculation for the spin-averaged angular correlation between two correlated electrons escaping the W(001) surface following the impact of one electron with the energy 17.2 eV. The coordinate system is as in Fig. 6.9, however grazing incidence of the incoming projectile electron is chosen ($\theta_0 = 88^\circ$ $\phi_0 = 0^\circ$). The two outgoing electrons possess asymptotically (at the detector) equal energies $E_1 = E_2 = 6$ eV. One electron is fixed at the angular position $\theta_1 = 47^\circ$ and $\phi_1 = 180^\circ$ (marked by the broken circle) whereas the angular correlation is scanned as a function of θ_2 and ϕ_2. The upper panel is the result of the calculations when $U_{\rm eff}$ is switched off in (6.35) whereas the lower panel shows the result of the LKKR calculations using the correlated two-particle LEED state. See also color figure on page 151.

barrier. In the vacuum half space the electrons feel the constant potential $V_{\text{vac}} = V_{\text{MT}} + |V_{\text{Or}}|$, where V_{Or} is the real part of the optical potential V_{O} and V_{MT} is the zero point of the MT potential. The surface-parallel atomic layers have a 2D periodicity and the potential is expanded in MT spheres. Usually, within the MT spheres the potential is spherically symmetric and otherwise constant in the interstial region. Each surface-parallel layer has a common 2D unit cell which contains, in the simplest case, one atom. Away from the surface-vacuum interface the solid is treated as in the bulk case (volume layers). In the LKKR approach one starts the calculations by computing at a definite energy the t matrix for the scattering of the electron from one MT sphere. Hereby one utilizes the effective one-particle Dirac equation. In a second step one employs the single-particle Green's function to calculate the multiple scattering from a 2D periodic layer as well as to account for the scattering between the layers. In this way one deduces the states of the electron in the semi-infinite solid with the appropriate boundary conditions.

6.4.3 Two-particle Energy Correlation in the Pair Emission from Tungsten

Figure 6.9 illustrates for a tungsten surface as an example, the correlated two-particle energy distribution at an impact energy of 10.6 eV. This means that electron pairs emitted from E_{F} have a maximal total kinetic energy of 6 eV. The calculations using two uncorrelated (antisymmetrized) LEED states for the escaping electrons (i. e., $Z_1 = 0 = Z_2$ in Eq. (6.36)) predict a minimum for the emission of equal energy ($E_1 = E_2$) correlated electron pairs from the vicinity of E_{F}, whereas maxima are anticipated for an asymmetric energy sharing, namely at $(E_1, E_2) = (4.8, 1.0)$ and $(1.0, 4.8)$. This is basically the high-energy behavior which we encountered and explained in the examples presented in the preceding sections. Obviously, this high-energy structure is not substantiated (cf. Fig. 6.9(b)) by the calculations including U_{eff}. In Fig. 6.9(b) the energy distribution possesses a maximum for equal energy sharing. The corresponding experimental data (cf. Fig. 6.9(c)) seem to endorse the conclusion of the calculation as to the importance of the electron–electron interaction in determining the two-particle emission intensity[5].

[5]We note that the substantial intensity in the experimental data shown in Fig. 6.9(c) for small energies of the two escaping electrons $E_j \sim 1$ eV is likely to stem from two accidentally time-correlated secondary electrons.

6.4.4 Angular Pair Correlation: Role of the Electron–Electron Interaction

Clear evidence of the importance of a strong correlation between the electrons is inferred from the two-particle angular distribution, i.e., by fixing one electron in a certain angular position with respect to the crystal orientation and scanning the two-particle coincidence intensity as a function of the angular position of the second electron, which is done in Fig. 6.10. As is obvious from the comparison of the results of the calculations with and without inclusion of the interelectronic interaction U_{eff} in (6.35), at low energies the angular pair correlation is dominated by the two-particle interaction. This is particularly true when the two electrons have similar wave vectors. For atomic systems we have already analyzed this feature in Chapter 3. Contrasting the atomic and the condensed matter case one can state the following: The crystal structure of a solid sample, i.e., the discrete translational symmetry imposes certain symmetry properties on the two-particle intensity (cf. Figs. 6.10 and 6.10) which are not encountered in the atomic case. The exchange and the correlation induced depression in the two-particle intensity around one electron when the two electrons are close in momentum space has a reduced extension in the condensed phase due to screening. Hence, a measurement of the type considered in Fig. 6.10 offers an insight into the strength of the material specific screening as well as into the local (on the scale of λ_{s}) properties of the electronic correlation. In other words, one can obtain information on the two-particle angular and energy correlation within a sphere of radius $\sim \lambda_{\text{s}}$. These statements remain valid regardless of the structural properties of the sample under study and, in particular, for disordered systems. This is because the wave vectors $\boldsymbol{k}_j,\ j = 0, 1, 2$ are measured externally. Hence, for their determination we do not need to rely on certain conservation laws, such as the wave-vector conservation law in the case of ordered materials.

7 Pair Emission from Alloys

Alloys are important physical systems from both a conceptual and a technological point of view. This is well documented by the extensive literature on this subject (see, e.g., [219, 220] and references therein). In this chapter we address a particular issue in this field, namely how electronic properties of alloys, and particularly those related to electronic correlations can be studied by means of electron-pair emission induced by the incidence of unpolarized electrons. Two general anticipations are notable regarding the potential of the correlated pair emission for the investigation of alloy properties: (i) Since the momenta of the incoming and the outgoing electrons are determined by experiment one can make use of the linear momentum conservation law to infer the amount of momentum that has been absorbed by the sample. In the high-energy regime this fact can be utilized to study the configurationally averaged spectral function of the sample, as shown explicitly below. (ii) As we demonstrated in previous chapters the electron–electron interaction is indeed screened with a screening length of the order of few times the interatomic separation. Hence, resolving the correlated two-particle emission intensity holds the promise of uncovering correlation effects related to short-range ordering in the alloys. This is particularly important for spin systems where spin-polarized beams and spin-sensitive detection techniques should be employed. As the momentum and the energy transfer are controlled experimentally one can, in principle, scan the crossover regime from distant to close collisions.

It should be mentioned however that the study of disordered systems by means of two-particle spectroscopy is still in a developing stage. Here we set up a simple theoretical model for the description of the correlated pair scattering from the surface of a substitutional binary alloy and point out the basic new physical phenomena that can be expected in this case.

Electronic Correlation Mapping: From Finite to Extended Systems. Jamal Berakdar

ISBN: 3-527-40350-7

7.1 Correlated Two-particle Scattering from Binary Substitutional Alloys

Let us consider the experiment of the kind sketched in Fig. 2.1, however the sample is a substitutional disordered binary alloy A_xB_{1-x} which consists of two type of atoms A and B. These components occur in the alloy with concentrations $c_A = x$ and $c_B = 1 - x$, respectively. We assume full randomness, disregarding any statistical correlation in the lattice sites occupation [219, 220]. In addition, effects related to positional disorder [219, 220] are not addressed here. For the concentration the following condition applies

$$c_A + c_B = 1 \qquad (7.1)$$

In what follows we employ the single-site approximation[1], i.e., the one-particle potential at the site j depends on the occupation of this site by the atoms A or B but not on the occupation of other sites, thus neglecting local environment effects. Furthermore, in what follows we study only electrons as a projectile and consider spin-averaged quantities. Generalization to other targets and exchange-induced effects is straightforward. The presence of disorder imposes a configurational average (hereafter indicated by $\langle \ldots \rangle$) of the cross section (6.30).

7.1.1 Pair Emission from Alloys in Transmission Mode

In a series of measurements (e.g., [4–6] and references therein) the experiment illustrated in Fig. 2.1 was performed in a high-energy transmission mode (i.e. $E_{1,2} > 1$ keV) and under the so-called Bethe-ridge conditions, i.e., in a scattering geometry, where the scattering cross section is dominated by a process of a binary encounter of the projectile electron with the sample electron (this scattering process is described by the amplitude T_{pe}, given by Eq. (6.2)). In this situation one may neglect surface-related effects and assume the scattering states of the vacuum electrons to be well described by plane waves. From Eq. (1.44) it follows then that

[1] It should be remarked that single-site approximations, such as the virtual crystal approximation, the averaged t matrix approximation or the coherent potential approximation, do not account for statistical fluctuations in the chemical composition and do not treat short-range order effects, such as disorder-induced localization of states and formation of magnetic moments [219, 220]. Influence of such phenomena on the two-particle correlation functions, which is of concern here, have not yet been studied.

the configurationally averaged cross section is expressible in the form[2] [192]

$$\begin{aligned}\langle\sigma(E_1,E_2,\Omega_1,\Omega_2)\rangle &= \frac{k_1 k_2}{(2\pi)^3 k_0}\,(\sigma(\Omega))_{\rm ee}\,\langle A^-(\boldsymbol{k},\varepsilon)\rangle\,,\\ \boldsymbol{k} &= \boldsymbol{k}_0-\boldsymbol{k}_1-\boldsymbol{k}_2 \end{aligned} \qquad (7.2)$$

Here $(\sigma(\Omega))_{\rm ee}$ is the (atomic) Mott cross section which does not depend on the spatial arrangement of the atoms. Measuring the cross section (7.2) for disordered systems thus yields direct information on the disorder-average of $\langle A^-(\boldsymbol{k},\varepsilon)\rangle$ of the sample's spectral function which, expressed in terms of the occupied single-particle orbitals, φ_{ϵ_i} reads

$$A^-(\varepsilon)=\sum_{i_{\rm occ}}|\varphi_{\epsilon_i}\rangle\langle\varphi_{\epsilon_i}|\delta(\varepsilon-\epsilon_i),\qquad \varepsilon=E_1+E_2-E_0. \qquad (7.3)$$

This finding highlights a decisive advantage in measuring $\langle A^-(\boldsymbol{k},\varepsilon)\rangle$ via two-particle spectroscopy as compared to ultraviolet photoemission spectroscopy in which the lack of translational symmetry in disordered systems hinders the direct determination of $\langle A^-(\boldsymbol{k},\varepsilon)\rangle$ since in this case $\boldsymbol{k}$ is not fixed by the wave-vector conservation law.

7.1.2 Pair Emission in Reflection Mode

In high-energy transmission mode we argued that only the amplitude $T_{\rm pe}$ is relevant. This situation changes in the case of the moderate and low-energy back-reflection mode, as considered in Chapter 6. Indeed, as we demonstrated in Fig. 6.7a in the back-reflection mode $T_{\rm pe}$ plays a subsidiary role. Hence, we suppress in this section the treatment of this amplitude and analyze in some detail the contribution of the amplitude $T_{\rm pc}$ to the two-particle emission cross section from alloys. Specifically we account explicitly for the scattering processes (b,d) depicted in Fig. 6.1 (the processes (a,c) enters in the calculations as scattering events mediated via initial binding). Furthermore, to account for the refraction of the emitted electrons at the vacuum–surface interface as well as for the damping of these electronic states inside the surface we utilize a re-normalized (single-particle) Green's function $\tilde{g}_0(E_j)$ instead of the free Green's function $g_0(E_j)$. Hence, the transition operator we are employing has the form

$$T_{\rm pc}=W_{\rm pc}\tilde{g}_0^+(E_1)U_{\rm eff}+U_{\rm eff}\tilde{g}_0^+(E_0)W_{\rm pc}, \qquad (7.4)$$

[2] Exchange effects are marginal at high energies and under the Bethe-ridge conditions.

which is a first-order expression in the potentials U_{eff} and W_{pc}. The spin-averaged cross section can then be expressed as

$$\sigma^{av}(E_1, \Omega_1, E_2, \Omega_2) = \frac{1}{4}\,\sigma(E_1, \Omega_1, E_2, \Omega_2)|_{S=0} + \frac{3}{4}\,\sigma(E_1, \Omega_1, E_2, \Omega_2)|_{S=1} \tag{7.5}$$

$$\sigma(E_1, \Omega_1, E_2, \Omega_2)|_S = c \sum_{i_{\mathrm{occ}}} \left\{ |\langle \Psi_{\boldsymbol{k}_1,\boldsymbol{k}_2}|T_{\mathrm{pc}}|\boldsymbol{k}_0, \varphi_{\epsilon_i}\rangle|^2 + (-)^S\, |\langle \Psi_{\boldsymbol{k}_2,\boldsymbol{k}_1}|T_{\mathrm{pc}}|\boldsymbol{k}_0, \varphi_{\epsilon_i}\rangle|^2 \right\} \times \delta(E_1 + E_2 - E_0 - \epsilon_i) \tag{7.6}$$

Here the sum runs over the occupied states (i_{occ}) and c is a kinematical constant. Utilizing Eq. (7.4) we can express Eq. (7.6) also in the form

$$\sigma(E_1, \Omega_1, E_2, \Omega_2) = c \sum_{i_{\mathrm{occ}}} |\langle \psi_{\boldsymbol{k}_2}|\mathcal{T}(\boldsymbol{k}_1, \boldsymbol{k}_0)|\varphi_{\epsilon_i}\rangle|^2 \delta(E_1 + E_2 - E_0 - \epsilon_i) \tag{7.7}$$

where $|\psi_{\boldsymbol{k}_2}\rangle$ is the scattering state of the electron emitted from the sample surface. The effective single-electron transition operator $\mathcal{T}(\boldsymbol{k}_1, \boldsymbol{k}_0)$ is defined as

$$\mathcal{T}(\boldsymbol{k}_1, \boldsymbol{k}_0) = \langle \boldsymbol{k}_1|W_{\mathrm{pc}}\tilde{g}_0^+(E_1)U_{\mathrm{eff}} + U_{\mathrm{eff}}\tilde{g}_0^+(E_0)W_{\mathrm{pc}}|\boldsymbol{k}_0\rangle \tag{7.8}$$

7.1.3 Scattering Potential from Binary Alloys

As is clear from Eqs. (7.4)–(7.8) the evaluation of the cross section entails the knowledge of the electron-alloy scattering potential W_{pc}. An expression for the potential W_{pc} is obtained if W_{pc} is expanded as a sum of MT potentials $V_j^{(s)}$ located at sites occupied by the atom $s = A$ or $s = B$ (cf. Eq. (6.4)). To describe randomness we introduce the occupation indices ξ_j. The random numbers take on the following values: $\xi_j = 1$ if the site j is occupied by an atom of type A and $\xi_j = 0$ if j is occupied by an atom of type B. In terms of ξ_j the on-site ionic potentials are given by

$$V_j^{\mathrm{ion}} = \xi_j V_j^{(\mathrm{A})} + (1 - \xi_j)V_j^{(\mathrm{B})} \tag{7.9}$$

The configurational average $\langle \xi_j \rangle$ of ξ_j is determined by the probability for the atom A to occupy the site j, i.e.

$$\langle \xi_j \rangle = x$$

where x is the concentration of A. In the so-called virtual crystal approximation (VCA) the single-site potential Eq. (7.9) is written in the form

$$V_j^{\mathrm{ion}} = xV_j^{(\mathrm{A})} + (1 - x)V_j^{(\mathrm{B})} \tag{7.10}$$

The average t-matrix approximation (ATA) goes beyond VCA in that multiple scattering events from $V_j^{(\mathrm{A/B})}$ are taken into account (one employs instead of potentials, the t-matrices $t_j^{(\mathrm{A/B})} = V_j^{(\mathrm{A/B})} + V_j^{(\mathrm{A/B})} g_0 t_j^{(\mathrm{A/B})}$, where g_0 is the free Green's function). If the potentials $V_j^{(\mathrm{A/B})}$ scatter weakly, the single-collision potential approximation (VCA) is obviously justified. This is particularly true at higher energies of the electrons. A well-established approach which is more accurate that ATA, is the coherent potential approximation (CPA). In this method the calculations using the ATA are performed self-consistently. This procedure is of particular importance when it comes to the description of ground state properties of alloys. All of these aforementioned methods have been implemented in standard band-structure computer codes [221], such as the linear muffin-tin orbitals, LMTO-CPA [222], and the Korringa–Kohn–Rostoker methods, KKR-CPA [223]. Here we will not discuss the manifestation of the presence of disorder in the electronic structure of alloys, such studies have long been conducted and are well documented in the literature (see, e.g., [219, 220] and references therein). The aim of this section is to investigate how the two- particle distribution functions are influenced by disorder. To obtain semi-analytical results we will employ the VCA which is reasonable if the electrons have high energies (compared to E_F) or/and in the low-concentration limit and/or if the strengths of the potentials $V_j^{(\mathrm{B})}$ and $V_j^{(\mathrm{A})}$ are of the same order.

7.1.4 Electronic States and Disorder Averaged Spectral Functions

Now we address the analytical determination of the configurationally averaged spectral function of the sample and the wave functions of the bound and the slow ejected electrons. We utilize for this purpose the model (6.14) and (6.15) which we employed in the preceding chapter, i.e., we view the sample as a semi-infinite metallic alloy extended in space in the negative z-direction. The electrons are confined by a step-like one-electron potential $V_2 = -V_\mathrm{O}\Theta(-z)$. Considering a (pure) surface built out of the atoms A or B only, then the potential depth reads

$$V_\mathrm{O}^\mathrm{C} = E_\mathrm{F}^\mathrm{C} + W^\mathrm{C} \quad (\mathrm{C} = \mathrm{A}, \mathrm{B})$$

Here E_F^C and W^C are the Fermi energies and the work functions of the respective materials.

Applying the VCA we obtain an expression for the depth of the confining potential in the binary alloy as

$$V_{\mathrm{O}} = xV_{\mathrm{O}}^{\mathrm{A}} + (1-x)V_{\mathrm{O}}^{\mathrm{B}} \tag{7.11}$$

As in Eq. (6.15) we determine now the electronic wave functions and the energy spectrum in the potential of the alloy. This yields

$$\begin{aligned} \varphi_{\boldsymbol{k}}(\boldsymbol{r}) &= \mathrm{e}^{\mathrm{i}\boldsymbol{k}_\parallel \boldsymbol{r}_\parallel} \left\{\Theta(z) B_{k_z} \mathrm{e}^{-\gamma z} + \Theta(-z)\left(\mathrm{e}^{\mathrm{i}k_z z} + A_{k_z}\mathrm{e}^{-\mathrm{i}k_z z}\right)\right\} \\ E_{\boldsymbol{k}} &= \frac{1}{2}(k_\parallel^2 + k_z^2) - V_{\mathrm{O}} \end{aligned} \tag{7.12}$$

where

$$A_{k_z} = \frac{k_z - \mathrm{i}\gamma}{k_z + \mathrm{i}\gamma}, \quad B_{k_z} = \frac{2k_z}{k_z + \mathrm{i}\gamma}, \quad \gamma = \sqrt{2V_{\mathrm{O}} - k_z^2} \text{ and } \sqrt{2V_{\mathrm{O}}} \geq k_z \geq 0$$

The Fermi energy is determined by the density n of the valence electrons in the alloy as follows

$$E_{\mathrm{F}} = (3\pi^2 n)^{2/3}/2, \quad n = N/\upsilon$$

The number of valence electrons in the alloy $N = xN_{\mathrm{A}} + (1-x)N_{\mathrm{B}}$ is given in terms of the valence electron numbers $N_{\mathrm{A,B}}$ in the virtual crystal composed of the atoms of type A and B. υ stands for the volume per atom in the alloy.

The configurationally averaged spectral function (7.3) is determined according to

$$\langle A^-(\varepsilon)\rangle = f(\varepsilon)\sum_{\boldsymbol{k}} |\varphi_{\boldsymbol{k}}\rangle\langle\varphi_{\boldsymbol{k}}|\delta(\varepsilon - E_{\boldsymbol{k}}) \tag{7.13}$$

with $f(\varepsilon)$ being the Fermi distribution function.

7.2 Incorporation of Damping of the Electronic States

As indicated by Eq. (7.7) for the calculation of the cross section one requires the scattering wave function of the propagating sample's electron with energy $E_2 = k_2^2/2$. In front of a surface such an electron will be accelerated due to the negative potential difference V_{Or} between the vacuum and the bulk of the surface. Inside the solid this state will be damped due to inelastic scattering events such as particle–hole pair and secondary electron creations. These damping processes lead to a finite inelastic mean-free path λ_{e} of the electron. The description of such damping processes goes beyond the scope of density functional theory and requires the utilization of dynamical theories capable of describing excited states. In practical calculations of scattering amplitudes, however, one usually accounts for the damping phenomenologically by adding to the confining potential (7.11) a small imaginary part (optical potential) [197] $\mathrm{i}V_{\mathrm{Oi}}(V_{\mathrm{O}} \gg V_{\mathrm{Oi}} > 0)$. The (real) potential difference V_{Or} at the vacuum/surface interface leads to a refraction of the electron wave and can be incorporated as a renormalization of the electron Green's function. This is done above by using $\tilde{g}_0$ instead of g_0 assuming that V_{Or} is a step function[3] at $z = 0$, $(V_{\mathrm{Or}} = V_{\mathrm{O}}\Theta(-z))$.

An exact expression for the optical potential

$$V_{\mathbf{O}}(\boldsymbol{r}_j; E_j) = V_{\mathrm{Or}}(\boldsymbol{r}_j; E_j) + \mathrm{i}V_{\mathrm{Oi}}(\boldsymbol{r}_j; E_j), \quad j = 0, 1, 2 \tag{7.14}$$

is unknown. An ad hoc solution to this problem is to assume V_{Or} be constant. As discussed in Section 1.5 the inherently inelastic electron–electron scattering processes play a minor role near E_{F}. Hence, the imaginary part can be approximated by the spatially constant, energy dependent form [224, 225]

$$V_{\mathrm{Oi}}(E_j) = a(E_j - E_F)^b \tag{7.15}$$

Here a, b are free parameters that are determined from a fitting to experimental data. Within this treatment of damping the electronic scattering states attains the form

$$\psi_{\boldsymbol{k}_2}(\boldsymbol{r}) = \mathrm{e}^{\mathrm{i}\boldsymbol{k}_{2,\parallel}\boldsymbol{r}_\parallel}\mathrm{e}^{-\alpha z}\left\{\Theta(z)\mathrm{e}^{\mathrm{i}k_{2,z}z} + \Theta(-z)\left(A_1\mathrm{e}^{\mathrm{i}k'_{2,z}z} + A_2\mathrm{e}^{-\mathrm{i}k'_{2,z}z}\right)\right\} \tag{7.16}$$

The functions occurring in this expression have the following form

$$A_1 = \frac{k'_{2,z} + k_{2,z} + \mathrm{i}\alpha}{2k'_{2,z}}, \quad A_2 = \frac{k'_{2,z} - k_{2,z} - \mathrm{i}\alpha}{2k'_{2,z}}, \quad \alpha = \frac{V_{\mathrm{Oi}}}{k'_{2,z}} \approx \frac{1}{\lambda_{\mathrm{e}}},$$
$$k'_{2,z} = \sqrt{\frac{k^2_{2,z} + 2V_{\mathrm{O}} + [(k^2_{2,z} + 2V_{\mathrm{O}})^2 - 4V^2_{\mathrm{Oi}}]^{1/2}}{2}} \approx \sqrt{k^2_{2,z} + 2V_{\mathrm{O}}} \tag{7.17}$$

The single-site ionic potential V_j^{ion} can be assumed to have the same functional form as in Eq. (6.11), however, as we have two types of atoms, we write for the ionic potential at the lattice site j in the alloy

$$V_s^{jj'} = \frac{Z_{j'}\exp(-\lambda_{j'}|\boldsymbol{r}_1 - \boldsymbol{R}_j|)}{|\boldsymbol{r}_1 - \boldsymbol{R}_j|}, \quad (j' = \mathrm{A}, \mathrm{B}) \tag{7.18}$$

[3] We note here that a number of variants for the functional dependence of V_{Or} do exist in the literature [226–229].

$\boldsymbol{R}_j$ specifies the spatial coordinate of the site j and the parameters $Z_{j'}$ and $\lambda_{j'}$ are determined, e.g., by fitting Eq. (7.18) to the one-electron muffin-tin potential obtained from self-consistent density-functional calculations within the local-density approximation, as done in the calculations presented below.

Furthermore Eq. (7.7) entails knowledge of the effective electron–electron scattering potential $\langle U_{\text{eff}}\rangle$. As demonstrated in the preceding chapter, the local static TF potential can serve this purpose reasonably well at energies above E_{F} and away from the plasma frequencies. Hence, we apply in what follows the TF potential for $\langle U_{\text{eff}}\rangle$.

For the evaluation of Eq. (7.7) we also need the configurational average of the renormalized single particle Green's function $\tilde{g}_0^+(\boldsymbol{p}', \boldsymbol{p}, E_l)$, $l = 0, 1$. In the momentum-space representation one derives that

$$\begin{aligned}
\langle\langle \boldsymbol{k}_l + \boldsymbol{p}' | \tilde{g}_0^+(E_l) | \boldsymbol{k}_l + \boldsymbol{p}\rangle\rangle &= \delta(\mathbf{p} - \boldsymbol{p}') \left\langle \tilde{g}_0^+(E_l)(\boldsymbol{k}_l + \mathbf{p})\right\rangle, \\
&= \frac{\delta(\boldsymbol{p} - \boldsymbol{p}')}{E_l' - \frac{(\boldsymbol{k}_{l,\parallel} + \boldsymbol{p}_\parallel)^2}{2} - \frac{(k'_{l,z} + p_z)^2}{2} + \mathrm{i}V_{\mathrm{Oi},l}}, \qquad (l = 0, 1), \\
E_l' &= E_l + V_{\mathrm{O}}; \quad k'_{l,z} = \sqrt{k_{l,z}^2 + 2V_{\mathrm{O}}} \qquad (7.19)
\end{aligned}$$

Physically E_l' and $k'_{l,z}$ are respectively the refracted electron's energy and the wave vector normal to the surface. The functions $V_{\mathrm{Oi},l}$ are the imaginary parts of the optical potentials experienced by the incoming and outgoing projectile electron.

7.3 Configurationally Averaged Cross Section

Within the MT approximation and making use of the expression (7.9) the effective transition operator (7.8) for the simultaneous two-particle emission from a random binary alloy is expanded as a sum over the lattice sites

$$\mathcal{T}(\boldsymbol{k}_1, \boldsymbol{k}_0) = \sum_j \mathcal{T}_j(\boldsymbol{k}_1, \boldsymbol{k}_0) \qquad (7.20)$$

$$\mathcal{T}_j(\boldsymbol{k}_1, \boldsymbol{k}_0) = \langle \boldsymbol{k}_1 | V_j^{\text{ion}} \tilde{g}_0^+(E_1) U_{\text{eff}} + U_{\text{eff}} \tilde{g}_0^+(E_0) V_j^{\text{ion}} | \boldsymbol{k}_0\rangle \qquad (7.21)$$

$$\mathcal{T}_j(\boldsymbol{k}_1, \boldsymbol{k}_0) = \xi_j \mathcal{T}_{j\text{A}}(\boldsymbol{k}_1, \boldsymbol{k}_0) + (1 - \xi_j)\mathcal{T}_{j\text{B}}(\boldsymbol{k}_1, \boldsymbol{k}_0) \qquad (7.22)$$

From these relations it follows that the cross section is expressible as

$$\sigma(E_1, \Omega_1, E_2, \Omega_2) = c \sum_{jj'} \langle \psi_{\boldsymbol{k}_2} | \mathcal{T}_j(\boldsymbol{k}_1, \boldsymbol{k}_0) A^-(\varepsilon) \mathcal{T}_{j'}^+(\boldsymbol{k}_1, \boldsymbol{k}_0) | \psi_{\boldsymbol{k}_2}\rangle \qquad (7.23)$$

To configurationally average over the cross section (7.23) we proceed as follows:

1. All two-electron on-site correlated terms are omitted when performing the configurational average. In this way on-site quantities associated with individual electrons are decoupled.

2. We neglect the fluctuation terms

$$\tilde{g}_0^+(E_l) - \langle \tilde{g}_0^+(E_l) \rangle = 0, \quad U_{\mathrm{eff}} - \langle U_{\mathrm{eff}} \rangle = 0; \quad l = 0, 1$$

and obtain for $\mathcal{T}_j(\boldsymbol{k}_1, \boldsymbol{k}_0)$ the following expression

$$\mathcal{T}_j(\boldsymbol{k}_1, \boldsymbol{k}_0) = \langle \boldsymbol{k}_1 | V_j^{\mathrm{ion}} \langle \tilde{g}_0^+(E_1) \rangle \langle U_{\mathrm{eff}} \rangle + \langle U_{\mathrm{eff}} \rangle \langle \tilde{g}_0^+(E_0) \rangle V_j^{\mathrm{ion}} | \boldsymbol{k}_0 \rangle \tag{7.24}$$

These simplifications allow a further advance in formulating the principal term in Eq. (7.23), namely

$$\begin{aligned}
\left\langle \psi_{\boldsymbol{k}_2} | \mathcal{T}_j A^-(\varepsilon) \mathcal{T}_{j'}^+ | \psi_{\boldsymbol{k}_2} \right\rangle &= \left\langle \langle \psi_{\boldsymbol{k}_2} | \langle \mathcal{T}_j(\boldsymbol{k}_1, \boldsymbol{k}_0) \rangle A^-(\varepsilon) \left\langle \mathcal{T}_{j'}^+(\boldsymbol{k}_1, \boldsymbol{k}_0) \right\rangle | \psi_{\boldsymbol{k}_2} \rangle \right\rangle \\
&+ \delta_{jj'} \Bigg\{ x \left\langle \langle \psi_{\boldsymbol{k}_2} | \mathcal{T}_{j\mathrm{A}}(\boldsymbol{k}_1, \boldsymbol{k}_0) A^-(\varepsilon) \mathcal{T}_{j\mathrm{A}}^+(\boldsymbol{k}_1, \boldsymbol{k}_0) | \psi_{\boldsymbol{k}_2} \rangle \right\rangle \\
&+ (1-x) \left\langle \langle \psi_{\boldsymbol{k}_2} | \mathcal{T}_{j\mathrm{B}}(\boldsymbol{k}_1, \boldsymbol{k}_0) A^-(\varepsilon) \mathcal{T}_{j\mathrm{B}}^+(\boldsymbol{k}_1, \boldsymbol{k}_0) | \psi_{\boldsymbol{k}_2} \rangle \right\rangle \\
&- \left\langle \langle \psi_{\boldsymbol{k}_2} | \langle \mathcal{T}_j(\boldsymbol{k}_1, \boldsymbol{k}_0) \rangle A^-(\varepsilon) \left\langle \mathcal{T}_j^+(\boldsymbol{k}_1, \boldsymbol{k}_0) \right\rangle | \psi_{\boldsymbol{k}_2} \rangle \right\rangle \Bigg\}
\end{aligned} \tag{7.25}$$

The configurationally averaged cross section is then expressible as the sum of two terms

$$\langle \sigma(E_1, \Omega_1, E_2, \Omega_2) \rangle = \langle \sigma^{\mathrm{coh}}(E_1, \Omega_1, E_2, \Omega_2) \rangle + \langle \sigma^{\mathrm{incoh}}(E_1, \Omega_1.E_2.\Omega_2) \rangle \tag{7.26}$$

The term $\langle \sigma^{\mathrm{coh}}(E_1, \Omega_1, E_2, \Omega_2) \rangle$ stands for the electron coherent scattering

$$\begin{aligned}
&\langle \sigma^{\mathrm{coh}}(E_1, \Omega_1, E_2, \Omega_2) \rangle \\
&\quad = c \sum_{jj'} \left\langle \left\langle \psi_{\boldsymbol{k}_2} | \langle \mathcal{T}_j(\boldsymbol{k}_1, \boldsymbol{k}_0) \rangle A^-(\varepsilon) \left\langle \mathcal{T}_{j'}^+(\boldsymbol{k}_1, \boldsymbol{k}_0) \right\rangle | \psi_{\boldsymbol{k}_2} \right\rangle \right\rangle
\end{aligned} \tag{7.27}$$

On the other hand the quantities

$$\begin{aligned}
\langle \sigma^{\mathrm{incoh}}(E_1, \Omega_1, E_2, \Omega_2) \rangle = c \sum_j \Bigg\{ & x \left\langle \langle \psi_{\boldsymbol{k}_2} | \mathcal{T}_{j\mathrm{A}}(\boldsymbol{k}_1, \boldsymbol{k}_0) A^-(\varepsilon) \mathcal{T}_{j\mathrm{A}}^+(\boldsymbol{k}_1, \boldsymbol{k}_0) | \psi_{\boldsymbol{k}_2} \rangle \right\rangle \\
&+ (1-x) \left\langle \langle \psi_{\boldsymbol{k}_2} | \mathcal{T}_{j\mathrm{B}}(\boldsymbol{k}_1, \boldsymbol{k}_0) A^-(\varepsilon) \mathcal{T}_{j\mathrm{B}}^+(\boldsymbol{k}_1, \boldsymbol{k}_0) | \psi_{\boldsymbol{k}_2} \rangle \right\rangle \\
&- \left\langle \langle \psi_{\boldsymbol{k}_2} | \langle \mathcal{T}_j(\boldsymbol{k}_1, \boldsymbol{k}_0) \rangle A^-(\varepsilon) \langle \mathcal{T}_j^+(\boldsymbol{k}_1, \boldsymbol{k}_0) \rangle | \psi_{\boldsymbol{k}_2} \rangle \right\rangle \Bigg\}
\end{aligned} \tag{7.28}$$

and

$$\left\langle \mathcal{T}_j(\boldsymbol{k}_1,\boldsymbol{k}_0)\right\rangle = x\mathcal{T}_{j\mathrm{A}}(\boldsymbol{k}_1,\boldsymbol{k}_0) + (1-x)\mathcal{T}_{j\mathrm{B}}(\boldsymbol{k}_1,\boldsymbol{k}_0) \tag{7.29}$$

describe incoherent scattering processes of the electrons from the alloy's potential.

7.3.1 Analytical Model for Configurationally Averaged Cross Section

To obtain explicit analytical and numerical results for the two-particle emission cross section from binary alloys we utilize the model presented in Section (7.1.4) for the treatment of the electronic properties of the disordered sample: Upon inserting Eq. (7.13) in Eqs. (7.27) and (7.28) we obtain

$$\left\langle \sigma^{\mathrm{coh}}(E_1,\Omega_1,E_2,\Omega_2)\right\rangle = cf(\varepsilon)\sum_{\boldsymbol{k}}\delta(\varepsilon-E_{\boldsymbol{k}})\left|\sum_j \langle\psi_{\boldsymbol{k}_2}|\left\langle\mathcal{T}_j(\boldsymbol{k}_1,\boldsymbol{k}_0)\right\rangle|\varphi_{\boldsymbol{k}}\rangle\right|^2 \tag{7.30}$$

$$\begin{aligned}\left\langle \sigma^{\mathrm{incoh}}(E_1,\Omega_1,E_2,\Omega_2)\right\rangle = cf(\varepsilon)\sum_{\boldsymbol{k}}\delta(\varepsilon-E_{\boldsymbol{k}})\sum_j\Big\{ x\left|\langle\psi_{\boldsymbol{k}_2}|\mathcal{T}_{j\mathrm{A}}(\boldsymbol{k}_1,\boldsymbol{k}_0)|\varphi_{\boldsymbol{k}}\rangle\right|^2 \\ +(1-x)\left|\langle\psi_{\boldsymbol{k}_2}|\mathcal{T}_{j\mathrm{B}}(\boldsymbol{k}_1,\boldsymbol{k}_0)|\varphi_{\boldsymbol{k}}\rangle\right|^2 - \left|\langle\psi_{\boldsymbol{k}_2}|\left\langle\mathcal{T}_j(\boldsymbol{k}_1,\boldsymbol{k}_0)\right\rangle|\varphi_{\boldsymbol{k}}\rangle\right|^2\Big\}\end{aligned} \tag{7.31}$$

The matrix elements in Eq. (7.30) are evaluated as follows

$$\begin{aligned}\langle\psi_{\boldsymbol{k}_2}|\left\langle\mathcal{T}_j(\boldsymbol{k}_1,\boldsymbol{k}_0)\right\rangle|\varphi_{\boldsymbol{k}}\rangle = \int \mathrm{d}\boldsymbol{Q}\mathrm{e}^{\mathrm{i}\mathbf{Q}\mathbf{R}_j}\left\langle\tilde{V}^{\mathrm{ion}}(\boldsymbol{Q})\right\rangle\left\langle\tilde{U}_{\mathrm{eff}}(\boldsymbol{q}-\boldsymbol{Q})\right\rangle \\ \times S_{\boldsymbol{k}_2,\boldsymbol{k}}(\boldsymbol{q}-\boldsymbol{Q})\Big[\langle\tilde{g}_0^+(\boldsymbol{k}_1+\boldsymbol{Q},E_1)\rangle + \langle\tilde{g}_0^+(\boldsymbol{k}_0-\boldsymbol{Q},E_0)\rangle\Big]\end{aligned} \tag{7.32}$$

$$\left\langle\tilde{V}^{\mathrm{ion}}(\boldsymbol{Q})\right\rangle = x\tilde{V}^{\mathrm{A}}(\boldsymbol{Q}) + (1-x)\tilde{V}^{\mathrm{B}}(\boldsymbol{Q}) = \frac{4\pi xZ_{\mathrm{A}}}{Q^2+\lambda_{\mathrm{A}}^2} + \frac{4\pi(1-x)Z_{\mathrm{B}}}{Q^2+\lambda_{\mathrm{B}}^2} \tag{7.33}$$

$$\left\langle\tilde{U}_{\mathrm{eff}}(\boldsymbol{q}-\boldsymbol{Q})\right\rangle = \frac{4\pi}{(\boldsymbol{q}-\boldsymbol{Q})^2+1/\lambda_s^2}, \quad \boldsymbol{q}=\boldsymbol{k}_0-\boldsymbol{k}_1 \tag{7.34}$$

$$\langle\tilde{g}_0^+(\boldsymbol{k}_1+\boldsymbol{Q},E_1)\rangle = \frac{1}{E_s' - \frac{(\boldsymbol{k}_{1,\parallel}+\boldsymbol{Q}_\parallel)^2}{2} - \frac{(k_{1,z}'+Q_z)^2}{2} + \mathrm{i}V_{\mathrm{Oi},1}} \tag{7.35}$$

$$\langle\tilde{g}_0^+(\boldsymbol{k}_0-\boldsymbol{Q},E_0)\rangle = \frac{1}{E_0' - \frac{(\boldsymbol{k}_{0,\parallel}-\boldsymbol{Q}_\parallel)^2}{2} - \frac{(k_{0,z}'-Q_z)^2}{2} + \mathrm{i}V_{\mathrm{Oi},0}} \tag{7.36}$$

$$S_{\boldsymbol{k}_2,\boldsymbol{k}}(\boldsymbol{q}-\boldsymbol{Q}) = \langle\psi_{\boldsymbol{k}_2}|\mathrm{e}^{\mathrm{i}(\boldsymbol{q}-\boldsymbol{Q})\boldsymbol{r}_2}|\varphi_{\boldsymbol{k}}\rangle = \delta(\boldsymbol{k}_\parallel+\boldsymbol{q}_\parallel-\boldsymbol{Q}_\parallel-\boldsymbol{k}_{2,\parallel})S_{k_{2,z},k_z}(q_z-Q_z) \tag{7.37}$$

$$S_{k_{2,z},k_z}(q_z - Q_z) = \frac{B_{k_z}}{\gamma + \mathrm{i}(k_{2,z} - q_z + Q_z)} + \frac{2\alpha + \mathrm{i}(k_z + k_{2,z} + q_z - Q_z)}{[\alpha + \mathrm{i}(k_z + q_z - Q_z)]^2 + k_{2,z}'^2} + \frac{A_{k_z}[2\alpha + \mathrm{i}(k_{2,z} + q_z - Q_z - k_z)]}{[\alpha + \mathrm{i}(q_z - Q_z - k_z)]^2 + k_{2,z}'^2} \tag{7.38}$$

Now we sum the matrix elements (7.32) over the lattice sites j and integrate over $\boldsymbol{Q}_\|$ to obtain

$$\langle \mathcal{T}_{\boldsymbol{k}_2,\boldsymbol{k}}(\boldsymbol{q}) \rangle = \sum_{l,\boldsymbol{g}_\|} \delta(\boldsymbol{k}_\| + \boldsymbol{q}_\| - \boldsymbol{g}_\| - \boldsymbol{k}_{2,\|}) \left\langle \mathcal{T}^{l,\boldsymbol{g}_\|}_{k_{2,z},k_z}(\boldsymbol{q}) \right\rangle \tag{7.39}$$

In this equation we introduced the functions

$$\begin{aligned}\left\langle \mathcal{T}^{l,\boldsymbol{g}_\|}_{k_{2,z},k_z}(\boldsymbol{q}) \right\rangle &= \mathrm{e}^{\mathrm{i}\boldsymbol{g}_\| \boldsymbol{t}^l_\|} \\ &\times \int \mathrm{d}Q_z \mathrm{e}^{\mathrm{i}Q_z z_l} S_{k_{2,z},k_z}(q_z - Q_z) \left\langle \tilde{V}^{\mathrm{ion}}(\boldsymbol{g}_\|, Q_z) \right\rangle \\ &\times \left\langle \tilde{U}_{\mathrm{eff}}(\boldsymbol{q}_\| - \boldsymbol{g}_\|, q_z - Q_z) \right\rangle \\ &\times \Big[\langle \tilde{g}_0^+(\boldsymbol{k}_{1,\|} + \boldsymbol{g}_\|, k_{1,z} + Q_z, E_1) \rangle \\ &\quad + \langle \tilde{g}_0^+(\boldsymbol{k}_{0,\|} - \boldsymbol{g}_\|, k_{0,z} - Q_z, E_0) \rangle \Big]\end{aligned} \tag{7.40}$$

In Eq. (7.39) the sums run over the lattice planes indexed by l and located at $z_l < 0$. z_0 coincides with the inter-plane spacing $-d_z/2$. The vector $\boldsymbol{t}^l_\|$ specifies the parallel displacement of the lth plane measured from a reference point in the surface. As in the preceding chapter one can employ residues calculus [68] and perform the integration occurring in Eq. (7.40) analytically. The result for (7.30) is

$$\begin{aligned}\langle \sigma^{\mathrm{coh}}(E_1, \Omega_1, E_2, \Omega_2) \rangle &= c f(\varepsilon) \sum_{\boldsymbol{g}_\|} \frac{\Theta[2(\varepsilon + V_{\mathrm{O}}) - (\boldsymbol{q}_\| - \boldsymbol{k}_{2,\|} - \boldsymbol{g}_\|)^2]}{k_z^0} \\ &\times \left| \sum_l \left\langle \mathcal{T}^{l,\boldsymbol{g}_\|}_{k_{2,z},k_z^0}(\boldsymbol{q}) \right\rangle \right|^2, \\ k_z^0 &= \sqrt{2(\varepsilon + V_{\mathrm{O}}) - (\boldsymbol{q}_\| - \boldsymbol{k}_{2,\|} - \boldsymbol{g}_\|)^2}\end{aligned} \tag{7.41}$$

The incoherent contribution to the cross section as given by Eq. (7.31) may be rewritten as

$$\langle \sigma^{\mathrm{incoh}}(E_1, \Omega_1, E_2, \Omega_2) \rangle = c f(\varepsilon) \sum_{\boldsymbol{k}} \delta(\varepsilon - E_{\boldsymbol{k}}) \mathbb{L} \tag{7.42}$$

$$\mathbb{L} = x \mathcal{L}^{\mathrm{A}}_{\boldsymbol{k}_2,\boldsymbol{k}}(\boldsymbol{k}_1, \boldsymbol{k}_0) + (1 - x) \mathcal{L}^{\mathrm{B}}_{\boldsymbol{k}_2,\boldsymbol{k}}(\boldsymbol{k}_1, \boldsymbol{k}_0) - \langle \mathcal{L}_{\boldsymbol{k}_2,\boldsymbol{k}}(\boldsymbol{k}_1, \boldsymbol{k}_0) \rangle \tag{7.43}$$

$$\begin{aligned}\mathcal{L}^{\mathrm{A}}_{\boldsymbol{k}_2,\boldsymbol{k}}(\boldsymbol{k}_1,\boldsymbol{k}_0) &= \sum_j |\langle\psi_{\boldsymbol{k}_2}|\mathcal{T}_{j\mathrm{A}}(\boldsymbol{k}_1,\boldsymbol{k}_0)|\varphi_{\boldsymbol{k}}\rangle|^2 \\ \mathcal{L}^{\mathrm{B}}_{\boldsymbol{k}_2,\boldsymbol{k}}(\boldsymbol{k}_1,\boldsymbol{k}_0) &= \sum_j |\langle\psi_{\boldsymbol{k}_2}|\mathcal{T}_{j\mathrm{B}}(\boldsymbol{k}_1,\boldsymbol{k}_0)|\varphi_{\boldsymbol{k}}\rangle|^2 \\ \langle\mathcal{L}(\boldsymbol{k}_1,\boldsymbol{k}_0)\rangle &= \sum_j |\langle\psi_{\boldsymbol{k}_2}|\left\langle\mathcal{T}_j(\boldsymbol{k}_1,\boldsymbol{k}_0)\right\rangle|\varphi_{\boldsymbol{k}}\rangle|^2 \end{aligned} \tag{7.44}$$

Performing steps similar as those which led us to Eq. (7.40) we find for Eq. (7.42) the expression

$$\begin{aligned}\left\langle\sigma^{\text{incoh}}(E_1,\Omega_1,E_2,\Omega_2)\right\rangle = f(\varepsilon)\int_0^{\pi/2} d\theta_k \sin\theta_k \int_0^{2\pi} d\varphi_k \sum_l \Bigg\{ x\left|\mathcal{T}^{l,\boldsymbol{k}_\parallel}_{k_{2,z},k_z}(\boldsymbol{q})\right|^2_{\mathrm{A}} \\ + (1-x)\left|\mathcal{T}^{l,\boldsymbol{k}_\parallel}_{k_{2,z},k_z}(\boldsymbol{q})\right|^2_{\mathrm{B}} - \left|\left\langle\mathcal{T}^{l,\boldsymbol{K}_\parallel}_{k_{2,z},k_z}(\boldsymbol{q})\right\rangle\right|^2 \Bigg\}\end{aligned} \tag{7.45}$$

Here $k = \sqrt{2(\varepsilon+V_{\mathrm{O}})}$, $k_z = k\sin\theta_k$ and the functions $\left|\mathcal{T}^{l,\boldsymbol{k}_\parallel}_{k_{2,z},k_z}(\boldsymbol{q})\right|_j$ are obtained from Eq. (7.40) by inserting the ionic potential corresponding to the atom $j = A, B$ whereas in evaluating $\left|\left\langle\mathcal{T}^{l,\boldsymbol{K}_\parallel}_{k_{2,z},k_z}(\boldsymbol{q})\right\rangle\right|^2$ one employs the potential $\left\langle\tilde{V}^{\text{ion}}\right\rangle$. The vector $\boldsymbol{K}_\parallel$ is given by $\boldsymbol{K}_\parallel = \boldsymbol{q}_\parallel + \boldsymbol{k}_\parallel - \boldsymbol{k}_{2,\parallel}$.

7.4 Numerical Results and Illustrations

For the preceding semi-analytical calculations to be viable one should apply the theory to sp bonded metals, such as aluminum. An example is shown in Fig. 7.1. Upon the impact of a spin-unpolarized 200 eV electron onto the (001) face of an aluminum sp-metal alloy two interacting electrons are ejected simultaneously with a fixed total kinetic energy $E_{\text{tot}} = E_0 - W$ (W is the work function) so that the initially bound conduction electron resides at the Fermi level. Figure 7.1 shows how E_{tot} is shared within the electron pair and how the energy sharing is changed for different materials. The structures in Fig. 7.1 in the energy region $0.5 > |(E_1 - E_2)/E_{\text{tot}}|$ and $1 \geq |(E_1 - E_2)/E_{\text{tot}}| > 0.5$ in the case of clean aluminum ($x = 1$) can be assigned to the electron pair diffraction with the reciprocal lattice vectors $\boldsymbol{g}_\parallel = (00)$ and $\boldsymbol{g}_\parallel = \pm(11)$. As discussed in Section (6.2.4), the width of the structure around $E_1 = E_2$ when it is transformed into a wave vector scale can be associated with $2k_{\mathrm{F}}$. The pair diffraction is not influenced by alloying provided the incoherent contribution (7.28) to the cross section is marginal. The coherent contribution (7.27) to the cross section is hardly

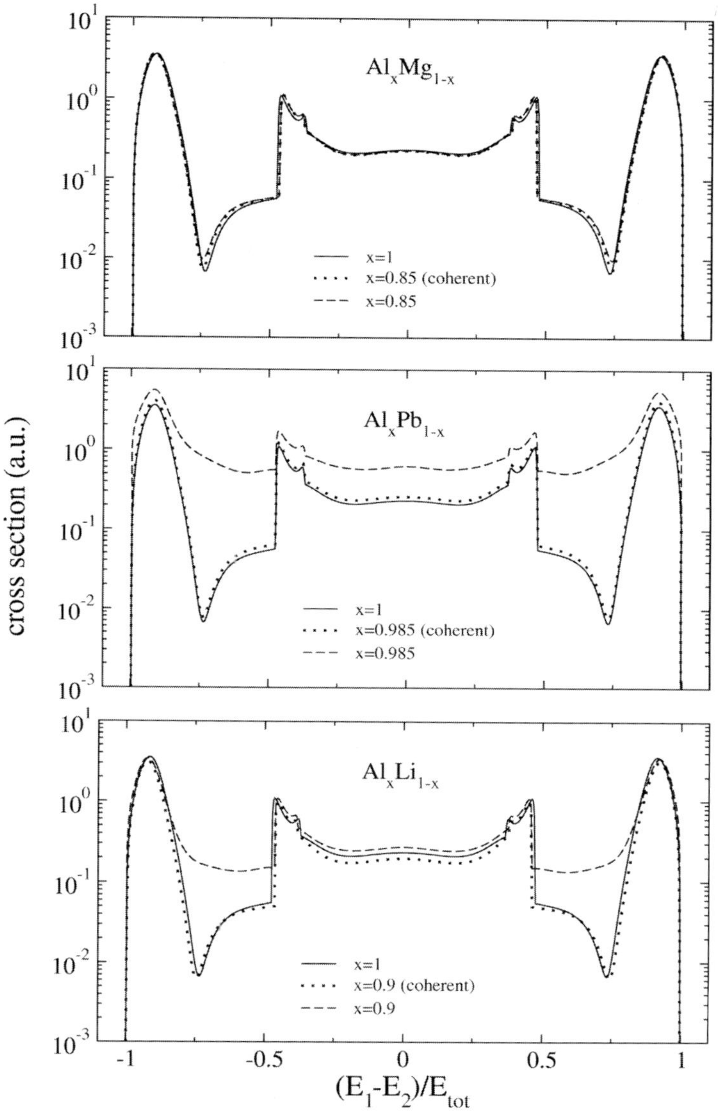

Figure 7.1: The spin-unresolved distribution of the fixed total kinetic energy $E_{\rm tot} = E_1 + E_2$ between two correlated electrons that escape from the (001) fcc face of the alloy indicated on the figure following the incident of 200 eV electrons. $E_{\rm tot}$ is chosen such that electron emission originates from the Fermi level ($E_{\rm tot} = E_0 - W$). $\boldsymbol{k}_0, \boldsymbol{k}_1$ and $\boldsymbol{k}_2$ are chosen to be in one plane, the x-z plane and the x-axis is along the [100] direction. The electron detectors are positioned at equal angles (45°) to the left and to the right of the normally incidence electron beam (which defines the z axis).The muffin-tin potentials are deduced from self-consistent density-functional calculations within the local-density approximation.

influenced by alloying (alloying results in a minor change in the k_F for $Al_{0.85}Mg_{0.15}$ and $Al_{0.9}Li_{0.1}$).

Figure 7.1 also illustrates the intuitive anticipation that if the on-site muffin-tin potentials of the alloy constituents are similar then the incoherent contribution (7.28) to the cross section (7.26) is subsidiary. This is indeed the case for $Al_{0.85}Mg_{0.15}$. In contrast, if these one-site potentials differ substantially, as in the case of $Al_{0.9}Li_{0.1}$ and $Al_{0.985}Pb_{0.015}$, the cross section for the two-particle emission is modified decisively.

In Fig. 7.1 we observe that for $Al_{0.9}Li_{0.1}$ and $Al_{0.985}Pb_{0.015}$ the enhanced incoherent backscattering results in qualitative changes in the energy-sharing distribution curves. In particular, alloying relaxes the conservation law for the surface-parallel wave vector of the center-of-mass of the electron pair. As a result, the pair diffraction becomes less pronounced (and eventually smeared out). More results and discussion, particularly of the effect of alloying on the angular correlation and the mapping of the alloys spectral function with the aid of two-particle spectroscopy can be found in [192, 230].

Color Figures

Chapter 5

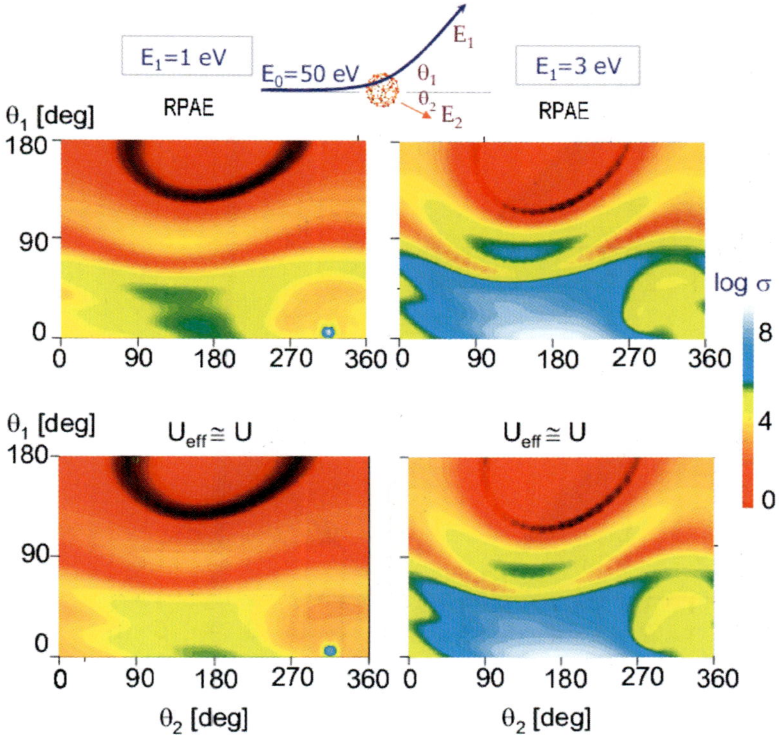

Figure 5.2: The angular dependence of the fully differential cross section for the emission of one electron from C_{60} with an energy of 1 eV (left panel) or 3 eV (right panel) following the impact of 50 eV electrons. A schematic of the scattering geometry is depicted. The upper part of the figure shows the RPAE calculations while the lower part shows the calculations without treating screening effects.

Electronic Correlation Mapping: From Finite to Extended Systems. Jamal Berakdar

ISBN: 3-527-40350-7

Chapter 6

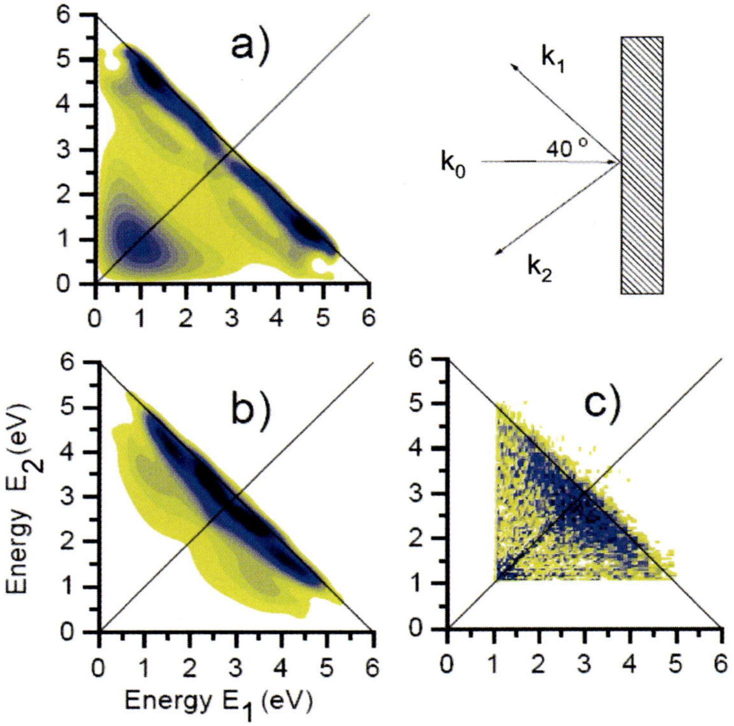

Figure 6.9: The two-particle energy distribution of two correlated spin-unresolved electrons emitted simultaneously upon the impact of a 10 eV spin-unpolarized electron onto the W(001) surface. $\boldsymbol{k}_0$ is aligned along the surface normal which defines the z axis. The x axis is chosen along the direction [100] of the sample. All momenta of the electrons $\boldsymbol{k}_0, \boldsymbol{k}_1, \boldsymbol{k}_2$ are chosen to be in one plane, the x-z plane. As shown schematically, the polar angles are fixed at $\theta_1 = \theta_2 = 40°$ and the azimuthal angles are $\phi_1 = 0$ and $\phi_2 = 180°$. Density plots show the results of the LKKR calculations when $U_{\text{eff}} \equiv 0$ in Eq. (6.35) (panel (a) whereas in panel (b) the full potential, as given by Eq. (6.35) is taken into account (under the approximation (6.38)). Experimental data are shown in panel (c). The energy regimes for which $\epsilon_i = E_{\text{F}}$ and $E_1 = E_2$ are indicated respectively by the solid counter-diagonal and the diagonal lines.

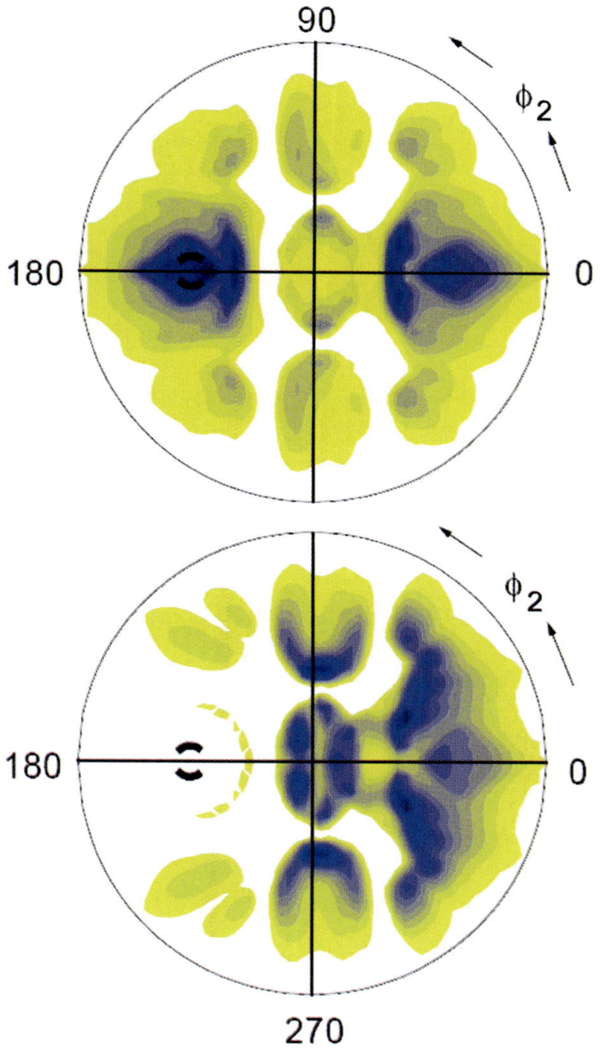

Figure 6.10: The results of the LKKR calculation for the spin-averaged angular correlation between two correlated electrons escaping the W(001) surface following the impact of one electron with the energy 17.2 eV. The coordinate system is as in Fig. 6.9, however grazing incidence of the incoming projectile electron is chosen ($\theta_0 = 88°$ $\phi_0 = 0°$). The two outgoing electrons possess asymptotically (at the detector) equal energies $E_1 = E_2 = 6$ eV. One electron is fixed at the angular position $\theta_1 = 47°$ and $\phi_1 = 180°$ (marked by the broken circle) whereas the angular correlation is scanned as a function of θ_2 and ϕ_2. The upper panel is the result of the calculations when $U_{\rm eff}$ is switched off in (6.35) whereas the lower panel shows the result of the LKKR calculations using the correlated two-particle LEED state.

Appendices

A Electronic States in a Periodic Potential

The quantum states ψ of an electron in a periodic, spin and time-independent crystal potential $V(\boldsymbol{r})$ are characterized by the Bloch energies $\epsilon_n(\boldsymbol{k})$, where $\boldsymbol{k}$ is the Bloch vector[1] and by the spin and the band index n. Due to the periodicity of the potential the Bloch theorem applies and the ansatz for the spatial part of the wave function

$$\psi_{n\boldsymbol{k}}(\boldsymbol{r}) = u_{n\boldsymbol{k}}(\boldsymbol{r})\mathrm{e}^{\mathrm{i}\boldsymbol{k}\cdot\boldsymbol{r}}$$

is appropriate. $\psi_{\boldsymbol{k}}(\boldsymbol{r})$ is normalized such that

$$\int \mathrm{d}^3\boldsymbol{r}\psi^*_{n\boldsymbol{k}}(\boldsymbol{r})\psi_{n'\boldsymbol{k}'}(\boldsymbol{r}) = \delta_{\boldsymbol{k},\boldsymbol{k}'}\delta_{nn'}, \quad \text{and} \quad \sum_{\boldsymbol{k}}^{1.\mathrm{BZ}} \psi^*_{n\boldsymbol{k}}(\boldsymbol{r})\psi_{n\boldsymbol{k}}(\boldsymbol{r}') = \delta(\boldsymbol{r}-\boldsymbol{r}')$$

where the $\boldsymbol{k}$ sum runs over the first Brillouin zone (BZ). The amplitude function $u_{\boldsymbol{k}}(\boldsymbol{r})$ possesses the periodicity of the underlying crystal potential and hence it is useful to expand it in Fourier components on a basis of plane waves of the translational vector of the reciprocal space $\boldsymbol{g}$, i.e.

$$u_{n\boldsymbol{k}}(\boldsymbol{r}) = \sum_{\boldsymbol{g}} c_{\boldsymbol{g}}(n\boldsymbol{k})\mathrm{e}^{\mathrm{i}\boldsymbol{g}\cdot\boldsymbol{r}} \tag{A.1}$$

Hence, $\psi_{n\boldsymbol{k}}$ is cast as

$$\psi_{n\boldsymbol{k}}(\boldsymbol{r}) = \sum_{\boldsymbol{g}} c_{\boldsymbol{g}}(n\boldsymbol{k})\mathrm{e}^{\mathrm{i}(\boldsymbol{g}+\boldsymbol{k})\cdot\boldsymbol{r}} \tag{A.2}$$

From this equation follows the meaning of $|c_{\boldsymbol{g}}(n\boldsymbol{k})|^2$ as a measure for the projection of $\psi_{n\boldsymbol{k}}(\boldsymbol{r})$ onto the plane wave characterized by $\boldsymbol{g}+\boldsymbol{k}$. The momentum space representation of (A.2) is[2]

$$\tilde{\psi}_{n\boldsymbol{k}}(\boldsymbol{p}) = (2\pi)^{3/2}\sum_{\boldsymbol{g}} c_{n\boldsymbol{g}}(\boldsymbol{k})\delta_{\boldsymbol{g}+\boldsymbol{k},\boldsymbol{p}} \tag{A.3}$$

[1] The expectation value of the momentum operator for an electron having the energy $\epsilon_n(\boldsymbol{k})$ is $\frac{m}{\hbar}\boldsymbol{\nabla}_{\boldsymbol{k}}\epsilon_n(\boldsymbol{k})$. From this expression one derives the group velocity of the wave packet (A.1). So, in a crystal, $\hbar\boldsymbol{k}$ is no longer the electron momentum expectation value. Therefore $\hbar\boldsymbol{k}$ is referred to as the crystal of pseudo momentum.

[2] Note that $\boldsymbol{k}$ is defined in the first Brillouin zone, whereas $\boldsymbol{p}$ scans the whole momentum space.

Electronic Correlation Mapping: From Finite to Extended Systems. Jamal Berakdar

ISBN: 3-527-40350-7

The orthogonality of $\psi_{\boldsymbol{k}}$ enforces that

$$\sum_{\boldsymbol{g}} c_{\boldsymbol{g}}(n\boldsymbol{k})c^*_{\boldsymbol{g}}(n'\boldsymbol{k}') = \delta_{\boldsymbol{k}\boldsymbol{k}'}\delta_{nn'}$$

Since the Bloch wave functions are periodic in the space spanned by $\boldsymbol{k}$ one can represent them as

$$\psi_{n\boldsymbol{k}}(\boldsymbol{r}) = \frac{1}{N}\sum_j a_n(\boldsymbol{R}_j, \boldsymbol{r})\mathrm{e}^{\mathrm{i}\boldsymbol{k}\cdot\boldsymbol{R}_j} \tag{A.4}$$

where the vectors $\boldsymbol{R}_j$ span the point lattice of the crystal and N is the number of lattice points. The expansion coefficients $a_n(\boldsymbol{R}_j, \boldsymbol{r})$ are called the Wannier functions. Inverting Eq. (A.4) we establish that

$$a_n(\boldsymbol{R}_j, \boldsymbol{r}) = \frac{1}{N}\sum_{vk}^{\mathbf{1.}\boldsymbol{BZ}} u_{n\boldsymbol{k}}(\boldsymbol{r})\mathrm{e}^{\mathrm{i}\boldsymbol{k}\cdot(\boldsymbol{r}-\boldsymbol{R}_j)}$$

Since $u_{n\boldsymbol{k}}(\boldsymbol{r})$ are lattice periodic $a_n(\boldsymbol{R}_j, \boldsymbol{r})$ depend only on $\boldsymbol{R}_j - \boldsymbol{r}$ and their amplitudes are peaked at the positions of the lattice sites.

The Wannier and the Bloch functions stand for two physical concepts of how to view electronic states in a periodic crystal potential: For Bloch waves the probability density for the electrons at a given point is the same for all equivalent lattice points which means that Bloch waves are *extended* states. In contrast Wannier functions are *localized* at the lattice sites.

The concept of the Wannier functions is instrumental in establishing the connection to the atomic nature of electronic states: When the interatomic separation is large the Wannier function $a_n(\boldsymbol{R}_j - \boldsymbol{r})$ tends increasingly to the atomic wave functions $\psi^{\text{at}}_\alpha(\boldsymbol{R}_j - \boldsymbol{r})$, where $\psi^{\text{at}}_\alpha(\boldsymbol{R}_j - \boldsymbol{r})$ satisfy the atomic Schrödinger equation at the site $\boldsymbol{R}_j$, i.e.,

$$\left[-\frac{\hbar^2}{2m}\nabla^2 - V_{\text{at}}(\boldsymbol{R}_j - \boldsymbol{r}) - \epsilon_{\text{at},\alpha}\right]\psi^{\text{at}}_\alpha(\boldsymbol{R}_j - \boldsymbol{r}) = 0$$

where V_{at} is the atomic potential. $\epsilon_{\text{at},\alpha}$ is the energy associated with the state characterized by the atomic quantum numbers α. An expression for the expansion coefficients in Eq. (A.4) derives if we approximate $a_n(\boldsymbol{R}_j, \boldsymbol{r})$ by $\psi^{\text{at}}_\alpha(\boldsymbol{R}_j - \boldsymbol{r})$. From the Schrödiger equation for $\psi_{n\boldsymbol{k}}(\boldsymbol{r})$ we can then deduce determining equations for the Bloch energies $\epsilon_n(\boldsymbol{k})$. For a crystal with cubic symmetry we find for example that

$$[\epsilon_\alpha(\boldsymbol{k}) - \epsilon_{\text{at},\alpha}] \sum_j \mathrm{e}^{\mathrm{i}\boldsymbol{k}\cdot\boldsymbol{R}_j} \left\langle \psi_\alpha^{\text{at}}(\boldsymbol{r}) | \psi_\alpha^{\text{at}}(\boldsymbol{R}_j - \boldsymbol{r}) \right\rangle$$
$$= \sum_j \mathrm{e}^{\mathrm{i}\boldsymbol{k}\cdot\boldsymbol{R}_j} \left\langle \psi_\alpha^{*at}(\boldsymbol{r}) | V(\boldsymbol{r}) - V_{\text{at}}(\boldsymbol{R}_j - \boldsymbol{r}) | \psi_\alpha^{\text{at}}(\boldsymbol{R}_j - \boldsymbol{r}) \right\rangle \quad \text{(A.5)}$$

If the wave functions on the neighboring atoms overlap weakly we can restrict our consideration to next neighbors (n.n.) terms in the sum and find for a cubic symmetry with a lattice constant a that

$$\epsilon_\alpha(\boldsymbol{k}) = \epsilon_{\text{at},\alpha} + C_\alpha + 2A_\alpha(\cos k_x a + \cos k_y a + \cos k_z a) \quad \text{(A.6)}$$

where

$$C_\alpha = \left\langle \psi_\alpha^{\text{at}}(\boldsymbol{r}) | V(\boldsymbol{r}) - V_{\text{at}}(\boldsymbol{r}) | \psi_\alpha^{\text{at}}(\boldsymbol{r}) \right\rangle$$

and

$$A_\alpha = \left\langle \psi_\alpha^{*\text{at}}(\boldsymbol{r}) | V(\boldsymbol{r}) - V_{\text{at}}(\boldsymbol{R}_{n.n.} - \boldsymbol{r}) | \psi_\alpha^{\text{at}}(\boldsymbol{R}_{n.n.} - \boldsymbol{r}) \right\rangle$$

Hence an atomic energy level $\epsilon_{\text{at},\alpha}$ evolves to a band of a width $12A_\alpha$ (each A_α varies in a range of $[-2A_\alpha, 2A_\alpha]$). The center of the band lies at $\epsilon_{\text{at},\alpha}$ displaced by C_α (which depends on the difference between the (free) atomic potential and the actual on-site potential in the lattice).

B Screening Within Linear Response Theory

Having outlined briefly some basic properties of electrons in solids the question we address now is how an electronic system responds to a (weak) perturbation δU switched on at $t = 0$. For instance δU can be the potential created by an approaching test charge (a projectile electron). In the event that the system responds in a linear manner to δU one can set up a theory that illuminates a number of important physical phenomena that occur when exciting electronic systems. Here some features and the main results of this theory are sketched with particular emphasis on the aspects that are of direct relevance to the materials presented in this treatise. A complete account can be found in standard textbooks on the subject, e.g. [125,146].

B.1 Kubo Formalism

The essence of linear response theory is that the system's response is proportional δU. The proportionality constant, i.e., the *linear response coefficient*, is then expressed in terms of a *correlation function* of the system in the equilibrium ensemble, i.e., when $\delta U = 0$. It is assumed that the equilibrium properties of the unperturbed system (which is described by the time-independent Hamiltonian H_0) are known.

The standard formulation relies on the Kubo formalism [231]. For the description of physical observables the time evolution of the statistical operator $\rho(t)$ is required. The evolution of $\rho(t)$ is governed by the Heisenberg equation

$$\mathrm{i}\partial_t \rho = [H_0 + \delta U, \rho] \tag{B.1}$$

Upon a canonical transformation $\tilde{\rho}(t) = S(t)\rho(t)S^\dagger(t)$ where $S(t) = \mathrm{e}^{\mathrm{i}H_0 t}$, Eq. (B.1) transforms as

$$\mathrm{i}\partial_t \tilde{\rho} = \left[\widetilde{\delta U}, \tilde{\rho}\right] \tag{B.2}$$

Electronic Correlation Mapping: From Finite to Extended Systems. Jamal Berakdar

ISBN: 3-527-40350-7

where $\widetilde{\delta U} = S(t)(\delta U)S^\dagger(t)$. The solution of the equation of motion (B.2) is formally given by

$$\tilde{\rho}(t) = \tilde{\rho}(0) - \mathrm{i}\int_0^t \left[\widetilde{\delta U}, \tilde{\rho}\right] \mathrm{d}t' \tag{B.3}$$

with $\tilde{\rho}(0) = \rho(t=0)$. The *leading order* iteration of (B.3) then reads

$$\begin{aligned} \tilde{\rho}(t) &\approx \rho(0) - \mathrm{i}\int_0^t \left[\widetilde{\delta U}, \rho(0)\right] \mathrm{d}t' \\ \rho(t) &\approx \rho(0) - \mathrm{i}S^\dagger(t)\left\{\int_0^t \left[\widetilde{\delta U}, \rho(0)\right] \mathrm{d}t'\right\} S(t) \end{aligned} \tag{B.4}$$

From these relations we can now evaluate the expectation value $\langle \mathcal{O} \rangle$ of an operator $\mathcal{O}(t)$ at temperature T and time t as $\langle \mathcal{O}(t) \rangle = \mathrm{tr}(\rho(t)\mathcal{O})$ which can be cast in terms of the density operator of the unperturbed system as

$$\langle \mathcal{O}(t) \rangle \approx \mathrm{tr}(\rho(0)\mathcal{O}) - \mathrm{i}\int_0^t \mathrm{tr}\left\{\rho(0)\left[\widetilde{\delta U}, \tilde{\mathcal{O}}(t)\right]\right\} \mathrm{d}t' \tag{B.5}$$

where $\tilde{\mathcal{O}} = S\mathcal{O}S^\dagger$ is the Heisenberg representation of $\mathcal{O}$. Hence, the change $\delta\langle \mathcal{O} \rangle$in the expectation value due to the application of δU is[1]

$$\delta\langle \mathcal{O}(t) \rangle \approx -\mathrm{i}\int_0^t \mathrm{tr}\left\{\rho(0)\left[\widetilde{\delta U}, \tilde{\mathcal{O}}(t)\right]\right\} \mathrm{d}t' \tag{B.8}$$

B.2 Density–density Correlation Functions

For a class of perturbations which are important in general and in particular to this present work δU acts on the electronic density. In this case δU is written as

$$\delta U = \int \mathrm{d}^3\boldsymbol{r}\, n(\boldsymbol{r}) U(\boldsymbol{r}, t) \tag{B.9}$$

where $n(\boldsymbol{r}) = \sum_{m_s} \psi^\dagger(\boldsymbol{r}, t)\psi(\boldsymbol{r}, t)$ is the density operator and $U(\boldsymbol{r}, t)$ is a (c-number) function characterizing the functional dependence of the (total) perturbation, i.e., the sum of the

[1]Typically δU has the form $\delta U = \int \mathrm{d}^3\boldsymbol{r}\mathcal{B}(\boldsymbol{r})\phi(\boldsymbol{r}, t)$, where $\mathcal{B}(\boldsymbol{r})$ is usually a single particle operator and $\phi(\boldsymbol{r}, t)$ is a function that depends on the type of the perturbation (cf. for example Eq. (B.99)). In this case $\delta\langle \mathcal{O} \rangle$ (i.e., Eq. (B.5)) is expressible in terms of a correlation function (which is related to the retarded two-particle Green's function (cf. for example Eq. (B.15)) as

$$\begin{aligned} \delta\langle \mathcal{O}(1) \rangle &\approx -\mathrm{i}\int_0^t \int \mathrm{tr}\left\{\rho(0)\left[\tilde{\mathcal{O}}(1), \tilde{\mathcal{B}}(1')\right]\right\} \phi(1')\, \mathrm{d}1' && \text{(B.6)} \\ &= \int G_{\mathcal{O}\mathcal{B}}(1, 1')\phi(1')\, \mathrm{d}1' && \text{(B.7)} \end{aligned}$$

Here $1 \equiv (\boldsymbol{r}, m_\mathrm{s}, t)$, where m_s denotes the spin variable.

external and induced perturbation. $\psi^\dagger(\boldsymbol{r},t)$ and $\psi(\boldsymbol{r},t)$ are creation and annihilation field operators.

The question of interest is how the charge density reacts to the applied perturbation. From Eqs. (B.7) and (B.8) we find to a leading order for the change in the density[2]

$$\langle \delta n(\boldsymbol{r},t)\rangle = -\mathrm{i}\int \mathrm{tr}\Big\{\rho(0)\big[\tilde{n}(1),\tilde{n}(1')\big]\Big\}U(1')d1' \tag{B.10}$$

$$= \int_{-\infty}^{\infty}\mathrm{d}t'\int \mathrm{d}^3\boldsymbol{r}'\chi^{\mathrm{R}}_{nn}(\boldsymbol{r},t;\boldsymbol{r}',t')U(\boldsymbol{r}',t') \tag{B.11}$$

and $$\langle \delta n(\boldsymbol{r},\omega)\rangle = \int \mathrm{d}^3\boldsymbol{r}'\chi^{\mathrm{R}}_{nn}(\boldsymbol{r},\boldsymbol{r}',\omega)U(\boldsymbol{r}',\omega) \tag{B.12}$$

Here we introduced the generalized susceptibility[3] in the time domain as $\chi^{\mathrm{R}}_{nn}(\boldsymbol{r},t;\boldsymbol{r}',t')$ and $\chi^{\mathrm{R}}_{nn}(\boldsymbol{r},\boldsymbol{r}',\omega))$ as its frequency-domain counterpart. $\chi^{\mathrm{R}}_{nn}(\boldsymbol{r},t;\boldsymbol{r}',t')$ is given by

$$\chi^{\mathrm{R}}_{nn}(\boldsymbol{r},t;\boldsymbol{r}',t') \equiv \chi^{\mathrm{R}}(\boldsymbol{r},t;\boldsymbol{r}',t') = -\mathrm{i}\Theta(t-t')\,\mathrm{tr}\Big\{\rho(0)\big[\tilde{n}(\boldsymbol{r},t),\tilde{n}(\boldsymbol{r}',t')\big]\Big\} \tag{B.14}$$

In a perturbative approach, to apply Wick's theorem one deals with the time-ordered version D of the retarded quantity D^{R} (B.13) and establishes then a relation between D and D^{R} (and the advanced correlation function D^{A}). Considering completely connected diagrams the zero-order D_0 and the first-order terms are readily evaluated. E.g. one finds that

$$D_0(\boldsymbol{r},t,\boldsymbol{r}',t') = -2\mathrm{i}G_0(\boldsymbol{r},t,\boldsymbol{r}',t')G_0(\boldsymbol{r}',t',\boldsymbol{r},t) \tag{B.15}$$

where $G_0(\boldsymbol{r}',t',\boldsymbol{r},t)$ is the (zero-order) single particle Green's function in the absence of perturbation ($D_0/\hbar$ is called the polarizibility). The first order term in the perturbation series has the algebraic structure

$$D = D_0 + D_0 u\frac{D}{\hbar} \tag{B.16}$$

where u is the naked Coulomb interaction. Introducing $\chi_0 = D_0/\hbar$ one then obtains the susceptibility in the random-phase approximation (RPA) as

$$\begin{aligned}\chi(\boldsymbol{r},t,\boldsymbol{r}',t') &= \chi_0(\boldsymbol{r},t,\boldsymbol{r}',t')\\ &+\int \mathrm{d}^3\boldsymbol{r}_1\mathrm{d}t_1\int \mathrm{d}^3\boldsymbol{r}_1'\mathrm{d}t_1'\chi_0(\boldsymbol{r},t,\boldsymbol{r}_1,t_1)u(\boldsymbol{r}_1,t_1,\boldsymbol{r}_1',t_1')\chi(\boldsymbol{r}_1',t_1',\boldsymbol{r}',t')\end{aligned} \tag{B.17}$$

---

[2] We assume that $U \equiv 0$ for $t < 0$. The Θ function in (B.14) then allows the extension of the time integral in (B.11) to the interval $(-\infty,\infty)$.

[3] In this work we use units in which $\hbar$ is unity. Hence, numerically $\chi^{\mathrm{R}}_{nn}(\boldsymbol{r},t;\boldsymbol{r}',t')$ is equal to the retarded correlation function or in our case it coincides with the retarded density–density correlation function

$$D^{\mathrm{R}}(\boldsymbol{r},t;\boldsymbol{r}',t') = \hbar\chi^{\mathrm{R}}_{nn}(\boldsymbol{r},t;\boldsymbol{r}',t') \tag{B.13}$$

Going over into the frequency space and taking into account that the Coulomb interaction is instantaneous (in the Coulomb gauge) we find that

$$\chi(\boldsymbol{r},\boldsymbol{r}',\omega) = \chi_0(\boldsymbol{r},\boldsymbol{r}',\omega) + \int \mathrm{d}^3\boldsymbol{r}_1\mathrm{d}^3\boldsymbol{r}_1'\chi_0(\boldsymbol{r},\boldsymbol{r}_1,\omega)u(\boldsymbol{r}_1,\boldsymbol{r}_1')\chi(\boldsymbol{r}_1',\boldsymbol{r}',\omega) \quad \text{(B.18)}$$

For a translationally invariant system χ takes on a simpler form in the wave vector space $\boldsymbol{q}$, e.g., for crystalline systems one can express χ in a basis set of Bloch states, as given by Eqs. (A.2)–(A.1) and finds that ($\tilde{u}(\boldsymbol{q})$ is the Fourier transform of the Coulomb potential)

$$\chi_{\boldsymbol{g},\boldsymbol{g}'}(\boldsymbol{q},\omega) = \chi_{0\boldsymbol{g},\boldsymbol{g}'}(\boldsymbol{q},\omega) + \sum_{\boldsymbol{g}_1} \chi_{0\boldsymbol{g},\boldsymbol{g}_1}(\boldsymbol{q},\omega)\tilde{u}(\boldsymbol{q}+\boldsymbol{g}_1)\chi_{\boldsymbol{g}_1,\boldsymbol{g}'}(\boldsymbol{q},\omega) \quad \text{(B.19)}$$

For a homogenous electron gas (jellium model) we then find the simpler relation

$$\chi(q,\omega) = \chi_0(q,\omega) + \chi_0(q,\omega)\tilde{u}(q)\chi(q,\omega) \quad \text{(B.20)}$$

$$\Rightarrow \chi(q,\omega) = \frac{\chi_0(q,\omega)}{1-\tilde{u}(q)\chi_0(q,\omega)} \quad \text{(B.21)}$$

The zero-order retarded susceptibility (or the retarded polarizibility) is called the Lindhard function which we will explore now in some detail.

C Lindhard Function

To put the above formulas and definitions in the context of this treatise let us consider an external charged particle impinging on a homogenous, translationally invariant electronic system. The "effective" distortion potential $U_{\rm eff}$ caused by this test charge is the sum of the external potential $U_{\rm ext}$ and the induced potential $U_{\rm ind}$. In a linear response approach one writes

$$\tilde{U}_{\rm eff}(\boldsymbol{q}) = \epsilon^{-1}(\boldsymbol{q})\tilde{U}_{\rm ext} \tag{C.1}$$

The proportionality coefficient is determined by the dielectric function ϵ. As stated by (B.7) the induced change in the density $\langle \delta n \rangle$ is proportional to $U_{\rm eff}$. This proportionality is governed by the linear response coefficient χ. On the other hand, Poisson's equation establishes a relationship between $U_{\rm ind}$ and $\langle \delta n \rangle$, namely

$$U_{\rm ind} = \frac{4\pi}{q^2}\langle \delta n \rangle$$

Therefore, the equation applies

$$U_{\rm eff} - U_{\rm ext} = U_{\rm eff}(1 - \epsilon(\boldsymbol{q})) = \tilde{u}(\boldsymbol{q})\chi U_{\rm eff}$$

from which we conclude that

$$\epsilon(\boldsymbol{q}) = 1 - \tilde{u}(\boldsymbol{q})\chi \tag{C.2}$$

From these straightforward arguments we see that χ is the central quantity in understanding how a charged particle couples to a polarizable electronic medium. For a homogenous, translationally invariant fermionic system the zero-order term χ_0 is readily evaluated in a plane-wave basis by calculating directly the ground state average in Eq. (B.14) [125]. The result is

$$\chi_0(\boldsymbol{q}, \omega) = \sum_{\boldsymbol{k} m_s} \frac{f(\xi_{\boldsymbol{k}}) - f(\xi_{\boldsymbol{k}+\boldsymbol{q}})}{\omega - (\xi_{\boldsymbol{k}+\boldsymbol{q}} - \xi_{\boldsymbol{k}}) + \mathrm{i}0^+} \tag{C.3}$$

Electronic Correlation Mapping: From Finite to Extended Systems. Jamal Berakdar

ISBN: 3-527-40350-7

where $f(\xi_{\boldsymbol{k}})$ is the Fermi–Dirac distribution at the energy $\xi_{\boldsymbol{k}}$ and the wave vector $\boldsymbol{k}$. So in a free gas the major contributions to the susceptibility stem from processes with the energy balance $\epsilon_{\boldsymbol{k}+\boldsymbol{q}} - \epsilon_{\boldsymbol{k}} - \omega = 0$. These are "particle–hole (p–h) excitations" where the energy ω that is transferred to the Fermi sea results in the excitation of a particle from below the Fermi surface (FS) (with the wave vector $\boldsymbol{k}$) to above the FS (with the wave vector $\boldsymbol{k}+\boldsymbol{q}$).

As evident from Eq. (B.18) the susceptibility of the free Fermion gas χ_0 ignores the effect of the long-range Coulomb interaction on the intermediate states. Hence it is in general a poor approximation.

C.1 Thomas–Fermi Approximation

Let us now explore[1] the static limit of Eq. (C.3). In the long-wavelength limit, i.e., for $q \to 0$ we find that for $T \ll E_{\mathrm{F}}$[2]

$$\chi_0 \to -2\sum_{\boldsymbol{k}} \frac{-\partial f}{\partial \epsilon_{\boldsymbol{k}}} \approx -DOS \int \mathrm{d}\xi_{\boldsymbol{k}} \frac{-\partial f}{\partial \epsilon_{\boldsymbol{k}}} = -DOS \tag{C.5}$$

Here DOS is the double-spin density of states. Hence, the static dielectric constant in the long-wavelength limit is [3]

$$\epsilon(q,0) \approx 1 + \frac{4\pi}{q^2} DOS = 1 + \frac{k_{\mathrm{TF}}^2}{q^2} \tag{C.6}$$

The Thomas–Fermi wave vector is given by $k_{\mathrm{TF}}^2 = 4\pi DOS$. In configuration space the effective potential U_{eff} is then exponentially screened with a (constant) screening length $\lambda_{\mathrm{s}} = k_{\mathrm{TF}}^{-1}$, i.e., U_{eff} has then the form

$$U_{\mathrm{eff}}(r) = \frac{\mathrm{e}^{-r/\lambda_{\mathrm{s}}}}{r} \tag{C.7}$$

[1] As is clear from Eq. (C.3) $\chi_0(\boldsymbol{q},\omega)$ has a real $\chi_0'(\boldsymbol{q},\omega)$ and an imaginary part

$$\chi_0''(\boldsymbol{q},\omega) \propto \sum_{\boldsymbol{k}m_{\mathrm{s}}} \delta\left(\omega - (\xi_{\boldsymbol{k}+\boldsymbol{q}} - \xi_{\boldsymbol{k}})\right)(f_{\boldsymbol{k}} - f_{\boldsymbol{k}+\boldsymbol{q}}) \tag{C.4}$$

which describes dissipation related to the creation of particle–hole pairs in the Fermi sea and sets the value of the density of states of particle–hole pair excitations. These are possible within a range of wave vectors spanning the Fermi surface. Outside this range χ_0 is purely real.

[2] This relation is an example of the general importance of the sharpness of the Fermi surface for the susceptibility.

[3] Recall that we use units in which the charge is measured in terms of electron charge.

C.2 Friedel Oscillations

The Thomas–Fermi (TF) theory is a long-wavelength approximation. It is valid for example in collisions with very small momentum transfer q. This is a particularly relevant case because the cross sections for scattering events with small q are in general dominant[4]. On the other hand TF is not capable of accounting for the response of the electronic system to a short-range perturbation. This case occurs in close collisions, or when an external charge is emersed into an electron gas. To obtain the effective potential in this case we evaluate the finite q dependence of the (static) charge susceptibility. The calculations show [125, 146] that the dielectric function has the form

$$\begin{aligned} \epsilon_{\text{Lindhard}} &= 1 - \tilde{u}(q)\chi_0(q, \omega = 0) = 1 - \frac{k_{\text{TF}}^2}{q^2} F\left(\frac{q}{2k_{\text{F}}}\right), \\ F(x) &= \frac{1-x^2}{4x} \log\left|\frac{1+x}{1-x}\right| + \frac{1}{2} \end{aligned} \tag{C.8}$$

The effective potential is then readily calculated to have the form

$$U_{\text{eff}}(\boldsymbol{r}) \approx \frac{\kappa^2}{(2+\kappa^2)^2} \frac{\cos 2k_{\text{F}} r}{r^3}, \tag{C.9}$$

where $\kappa = k_{\text{TF}}/(2k_{\text{F}})$.

C.3 Plasmon Excitations

Inspecting the structure of the dielectric function we see that when $\tilde{u}(\boldsymbol{q})\chi_0(\boldsymbol{q}, \omega) = 1$ the dielectric function vanishes, meaning that in this case the system response to an applied perturbation is particularly strong. To determine where this happens in the energy and in the wave vector space we note that χ_0 possesses simple poles at the particle–hole excitation energies

$$\pm\omega_{\boldsymbol{q}}^{\text{ph}}(\boldsymbol{k}) = \epsilon_{\boldsymbol{k}+\boldsymbol{q}} - \epsilon_{\boldsymbol{k}}$$

where $|\boldsymbol{k} + \boldsymbol{q}|$ is above the Fermi surface and $|\boldsymbol{k}|$ always below. This means χ_0 changes sign and diverges at $\omega_{\boldsymbol{q}}^{\text{ph}}$ and hence there will be a solution to $\tilde{u}(\boldsymbol{q})\chi_0(\boldsymbol{q}, \omega) = 1$ at each excitation energy. At $T = 0$ and for a given q there is a maximum frequency ω_{max} above which no particle–hole excitations are possible. The largest excitation takes place when $k = k_{\text{F}}$ and $\boldsymbol{k} \parallel \boldsymbol{q}$, so that for a parabolic dispersion we find

$$\omega_{\text{max}} = \frac{qk_{\text{F}}}{m} + \frac{q^2}{2m}$$

[4]This statement is based on the the form factor of the TF potential.

For $\omega > \omega_{\text{max}}$ one finds yet another solution to the equation $\Re\left[\tilde{u}(\boldsymbol{q})\chi_0(\boldsymbol{q},\omega)\right] = 1$. This is accomplished by making use of the small q expansion (at $T = 0$) of the susceptibility which yields

$$\begin{aligned}
\omega &= \omega_{\text{pl}}\left(1 + \mathcal{O}\left(\frac{q}{k_{\text{F}}}\right)^2 + \cdots\right)_{\text{max}} \\
\omega_{\text{pl}} &= \left(\frac{4\pi n e^2}{m}\right)^2 \qquad \text{(C.10)}
\end{aligned}$$

Here we explicitly included the electron charge e. ω_{pl} corresponds to the plasma frequency at which the particles are "collectively" set in motion.

These findings mean that there is a specific region in the (q,ω) space where the excitations of the electron gas are particle–hole (p–h) transitions. For $q \to 0$ plasma excitations occur at the finite frequency ω_{p}. This collective excitation mode disperses according to Eq. (C.10) and merges eventually into the p–h excitation region in which case the plasmon is strongly damped due to the decay into a p–h pair.

This physical picture depends strongly on the dimensionality of the system. E.g., for the two-dimensional electron gas F. Stern derived [232] that

$$\chi_{2D}(\boldsymbol{q}) = \begin{cases} -\frac{2me^2}{h} & \text{if} \quad q \leq 2k_{\text{F}} \\ -\frac{2me^2}{h}\left(1 - \sqrt{1 - 2k_{\text{F}}/q}\right) & \text{if} \quad q > 2k_{\text{F}} \end{cases} \qquad \text{(C.11)}$$

which leads to a $(2k_{\text{F}}r)^{-2}$ decay of the Friedel oscillation. For the plasma dispersion in the long-wavelength limit one finds (v_{F} is the Fermi velocity)

$$\omega_p^2 = \frac{2\pi n e^2}{m} q + \frac{3}{4} q^2 v_{\text{F}}^2 + \mathcal{O}(q^3) \qquad \text{(C.12)}$$

This indicates the absence of an energy gap (for $q \to 0$), in contrast to the three-dimensional case. In one dimension the susceptibility diverges at $q = 2k_{\text{F}}$, namely as

$$\chi_{1D}(q) = -\frac{e^2}{\pi q} \ln \frac{q + 2k_{\text{F}}}{q - 2k_{\text{F}}}$$

D Dynamic Structure Factor and the Pair-distribution Function

Let us consider a system which is described by the time-independent Hamiltonian H_0 and consists of N electrons in the volume V. The particle density operator $\hat{n}(\boldsymbol{r})$ is defined as the single-particle operator (in the sense of second quantization)

$$\hat{n}(\boldsymbol{r}) = \sum_{i=1}^{N} \delta(\boldsymbol{r} - \hat{\boldsymbol{r}}_i) \qquad \text{(D.1)}$$

Here $\boldsymbol{r}$ is a variable whereas $\hat{\boldsymbol{r}}_i$ is the electron position operator. In terms of creation ($a^\dagger_{\boldsymbol{k},m_\mathrm{s}}$) and annihilation ($a_{\boldsymbol{k},m_\mathrm{s}}$) operators of the single particle Bloch states, given by Eq. (A.2), $\hat{n}(\boldsymbol{r})$ read

$$\begin{aligned}
\hat{n}(\boldsymbol{r}) &= \sum_{\boldsymbol{k},\boldsymbol{k}',m_\mathrm{s},m_{\mathrm{s}'}} \langle \boldsymbol{k} m_\mathrm{s} | \delta(\boldsymbol{r} - \hat{\boldsymbol{r}}' | \boldsymbol{k}' m_{\mathrm{s}'} \rangle a^\dagger_{\boldsymbol{k}m_\mathrm{s}} a_{\boldsymbol{k}' m_{\mathrm{s}'}} \\
&= \sum_{\boldsymbol{k},\boldsymbol{k}',m_\mathrm{s},m_{\mathrm{s}'}} \delta_{m_\mathrm{s} m_{\mathrm{s}'}} \psi^*_{\boldsymbol{k}}(\boldsymbol{r}) \psi_{\boldsymbol{k}'}(\boldsymbol{r}) a^\dagger_{\boldsymbol{k}m_\mathrm{s}} a_{\boldsymbol{k}' m_{\mathrm{s}'}} \qquad \text{(D.2)}
\end{aligned}$$

m_s stands for the electron spin. For a homogenous, translationally invariant system Eq. (D.2) simplifies further to the form

$$\hat{n}(\boldsymbol{r}) = \frac{1}{V} \sum_{\boldsymbol{q}} \left\{ \sum_{\boldsymbol{k},m_\mathrm{s}} a^\dagger_{\boldsymbol{k}m_\mathrm{s}} a_{\boldsymbol{k}+\boldsymbol{q}m_\mathrm{s}} \right\} \mathrm{e}^{\mathrm{i}\boldsymbol{q}\cdot\boldsymbol{k}} = \frac{1}{V} \sum_{\boldsymbol{q}} \hat{n}_{\boldsymbol{q}} \mathrm{e}^{\mathrm{i}\boldsymbol{q}\cdot\boldsymbol{k}} \qquad \text{(D.3)}$$

This equation reveals the structure of the Fourier components $\hat{n}_{\boldsymbol{q}} = \sum_{\boldsymbol{k},m_\mathrm{s}} a^\dagger_{\boldsymbol{k}m_\mathrm{s}} a_{\boldsymbol{k}+\boldsymbol{q}m_\mathrm{s}}$ of the electron density operator . Note that $\hat{n}^\dagger_{\boldsymbol{q}} = \hat{n}_{-\boldsymbol{q}}$ and $\hat{n}_{\boldsymbol{q}=0} = \hat{N}$, where $\hat{N}$ is the particle number operator.

The particle density correlation is given by the relation

$$G(\boldsymbol{r},t) = \frac{1}{N} \int \mathrm{d}^3\boldsymbol{r}' \langle \hat{n}(\boldsymbol{r}' - \boldsymbol{r}, t=0)\, \hat{n}(\boldsymbol{r}',t) \rangle_{\text{ground state}} \qquad \text{(D.4)}$$

Here the time dependence of $\hat{n}(t)$ follows from the relation $\hat{n}(t) = \mathrm{e}^{\mathrm{i}H_0 t} \hat{n} \mathrm{e}^{-\mathrm{i}H_0 t}$.

Electronic Correlation Mapping: From Finite to Extended Systems. Jamal Berakdar

ISBN: 3-527-40350-7

Inserting Eq. (D.3) in Eq. (D.4) we find that

$$G(\boldsymbol{r},t) = \frac{1}{NV}\sum_{\boldsymbol{q}} \mathrm{e}^{-\mathrm{i}\boldsymbol{q}\cdot\boldsymbol{r}} \, \langle \hat{n}_{\boldsymbol{q}}(t=0)\, \hat{n}_{-\boldsymbol{q}}(t)\rangle_{\text{ground state}} \tag{D.5}$$

From Eq. (D.4) we infer that G has the meaning of the conditional probability for finding a particle at the position $\boldsymbol{r}$ and time t if another particle is present at the position $\boldsymbol{r} = 0$ and $t = 0$. Note that this correlation depends only on the relative coordinates $\boldsymbol{r}$ and t because we assumed the system to be spatially and temporally homogenous. Indeed G is related to the pair distribution function $g(r)$ which is a measure of the probability of finding, at a certain time, two particles within a relative distance r. Specifically, $g(\boldsymbol{r})$ is defined as

$$ng(\boldsymbol{r}) = \frac{1}{N}\sum_{ij,i\neq j} \langle \delta(\boldsymbol{r} + \hat{\boldsymbol{r}}_i(0) - \hat{\boldsymbol{r}}_j(0))\rangle, \quad \bar{n} := N/V \tag{D.6}$$

It is straightforward to establish that

$$\begin{aligned} G(\boldsymbol{r},0) &= \delta(\boldsymbol{r}) + \frac{1}{N}\sum_{ij,i\neq j} \langle \delta(\boldsymbol{r} + \hat{\boldsymbol{r}}_i(0) - \hat{\boldsymbol{r}}_j(0))\rangle, \\ &\Rightarrow \quad g(\boldsymbol{r}) = [G(\boldsymbol{r},0) - \delta(\boldsymbol{r})]/\bar{n} \end{aligned} \tag{D.7}$$

The dynamical structure factor $S(\boldsymbol{q},\omega)$ at the wave vector $\boldsymbol{q}$ and frequency ω is defined as the time and spatial Fourier transform of the particle density correlation $G(\boldsymbol{r},t)$, i.e.

$$S(\boldsymbol{q},\omega) = \int \mathrm{d}^3\boldsymbol{r} \int_{-\infty}^{\infty} \mathrm{d}t\, G(\boldsymbol{r},t)\mathrm{e}^{\mathrm{i}(\boldsymbol{q}\cdot\boldsymbol{r} - \omega t)} \tag{D.8}$$

Inserting in this equation the expression (D.5) for $G(\boldsymbol{r},t)$ we find

$$\begin{aligned} S(\boldsymbol{q},\omega) &= \frac{1}{NV}\sum_{\boldsymbol{q}'} \int \mathrm{d}^3\boldsymbol{r} \int_{-\infty}^{\infty} \mathrm{d}t\, \mathrm{e}^{\mathrm{i}(\boldsymbol{q}-\boldsymbol{q}')\cdot\boldsymbol{r}} \mathrm{e}^{-\mathrm{i}\omega t} \langle \hat{n}_{\boldsymbol{q}}(t=0)\, \hat{n}_{-\boldsymbol{q}'}(t)\rangle_{\text{ground state}} \\ &= \frac{1}{N}\int_{-\infty}^{\infty} \mathrm{d}t\, \mathrm{e}^{-\mathrm{i}\omega t} \langle \hat{n}_{\boldsymbol{q}}(t=0)\, \hat{n}_{-\boldsymbol{q}}(t)\rangle_{\text{ground state}} \end{aligned} \tag{D.9}$$

The static structure factor $S(\boldsymbol{q})$ is then deduced from Eq. (D.9) as

$$S(\boldsymbol{q}) := \int_{-\infty}^{\infty} \mathrm{d}\omega S(\boldsymbol{q},\omega) \tag{D.10}$$

$$S(\boldsymbol{q}) = \frac{2\pi}{N} \langle \hat{n}_{\boldsymbol{q}}(t=0)\, \hat{n}_{-\boldsymbol{q}}(t)\rangle_{\text{ground state}} \tag{D.11}$$

The zero-temperature average in Eq. (D.9) over the quantum mechanical ground state $|\epsilon_0\rangle$ is readily performed by utilizing a complete set $I = \sum_n |\epsilon_n\rangle\langle\epsilon_n|$ of energy eigenstates of H_0

and rewriting Eq. (D.9) as follows

$$
\begin{aligned}
S(\boldsymbol{q},\omega) &= \frac{1}{N}\int_{-\infty}^{\infty} \mathrm{d}t\,\mathrm{e}^{-\mathrm{i}\omega t}\sum_n \langle\epsilon_0|\hat{n}_{\boldsymbol{q}}(t=0)|\epsilon_n\rangle\langle\epsilon_n|\hat{n}_{-\boldsymbol{q}}(t)|\epsilon_0\rangle \\
&= \frac{2\pi}{N}\sum_n \langle\epsilon_0|\hat{n}_{\boldsymbol{q}}|\epsilon_n\rangle\langle\epsilon_n|\hat{n}_{-\boldsymbol{q}}|\epsilon_0\rangle\delta(\omega-(\epsilon_n-\epsilon_0)/\hbar) \\
S(\boldsymbol{q},\omega) &= \frac{2\pi}{N}\sum_n |\langle\epsilon_0|\hat{n}_{\boldsymbol{q}}|\epsilon_n\rangle|^2\delta(\omega-(\epsilon_n-\epsilon_0)/\hbar),\quad \hat{n}^\dagger_{\boldsymbol{q}} = \sum_{\boldsymbol{k}m_\mathrm{s}} a^\dagger_{\boldsymbol{k}+\boldsymbol{q}m_\mathrm{s}}a_{\boldsymbol{k}m_\mathrm{s}}
\end{aligned}
\tag{D.12}
$$

Here we make use of the fact that $\hat{n}(t) = \mathrm{e}^{\mathrm{i}H_0t}\hat{n}\mathrm{e}^{-\mathrm{i}H_0t}$ and perform the time Fourier transform and also account for $\hat{n}^\dagger_{\boldsymbol{q}} = \hat{n}_{-\boldsymbol{q}}$.

D.1 Excitation Processes and the Dynamical Structure Factor

Equation (D.12) elucidates the physical meaning of $S(\boldsymbol{q},\omega)$. It is related to the cross section of a process in which an external perturbation that couples to the (ground state) density transfers a wave vector $\boldsymbol{q}$ and energy ω. This results in particle–hole excitations. In total, however, the system remains neutral, in contrast to ionizing processes.

To give a concrete example, let us consider a fast (several keV energy) electron with an initial momentum $\boldsymbol{k}_0$ traversing a sample in its ground state. The electron beam transfers to the target charge density a small amount of energy (small compared to the work function) and wave vector and exits with the momentum $\boldsymbol{k}_1$ (cf. Fig. 1.1). The differential cross section for this process $\frac{\mathrm{d}\sigma_\mathrm{eels}}{\mathrm{d}\Omega_{\boldsymbol{q}}\mathrm{d}\omega}$ is connected with the dynamical structure factor $S(\boldsymbol{q}=\boldsymbol{k}_0-\boldsymbol{k}_1,\omega)$ by the relation [41]

$$
\frac{\mathrm{d}\sigma_\mathrm{eels}}{\mathrm{d}\Omega_{\boldsymbol{q}}\mathrm{d}\omega} \propto \left(\frac{\mathrm{d}\sigma}{\mathrm{d}\Omega_{\boldsymbol{q}}\mathrm{d}\omega}\right)_\mathrm{Rutherford} S(\boldsymbol{q},\omega)
\tag{D.13}
$$

Here $\left(\frac{\mathrm{d}\sigma}{\mathrm{d}\Omega_{\boldsymbol{q}}\mathrm{d}\omega}\right)_\mathrm{Rutherford}$ is the Rutherford cross section. Equations (D.4)–(D.8) also make clear that measuring the cross section (D.13) provides information on the electron pair distribution function. It is also obvious from the preceding section that the cross section (D.13) must be related to the charge-density-density correlation function and the longitudinal dielectric function ϵ. Indeed using the dissipation-fluctuation theorem [125] we can express the particle density correlation function (and hence $S(\boldsymbol{q},\omega)$ in terms of the imaginary part of the charge susceptibility $\chi''(\boldsymbol{q},\omega)$ (cf. Eq. (C.4)), i.e.

$$S(\boldsymbol{q},\omega) = \frac{V}{\pi} \frac{\chi''(\boldsymbol{q},\omega)}{1 - \mathrm{e}^{-\hbar\omega/(k_\mathrm{B}T)}}$$

where k_B is the Boltzmann constant and T is the temperature. Using the relation between χ and the longitudinal dielectric function of the electron gas ($\epsilon^{-1} = 1 - \tilde{u}(\boldsymbol{q})\chi(\boldsymbol{q},\omega)$) one finds then that

$$\chi''(\boldsymbol{q},\omega) = -\frac{1}{\tilde{u}(\boldsymbol{q})} \Im \frac{1}{\epsilon(\boldsymbol{q},\omega + \mathrm{i}0^+)}$$

Hence, at $T = 0$ we conclude that

$$\frac{\mathrm{d}\sigma_\mathrm{eels}}{\mathrm{d}\Omega_{\boldsymbol{q}}\mathrm{d}\omega} \propto \frac{k_1}{k_0} \tilde{u}(\boldsymbol{q}) \Im \left[-\frac{1}{\epsilon(\boldsymbol{q},\omega + \mathrm{i}0^+)} \right] \tag{D.14}$$

Hence, zero points in the longitudinal dielectric functions results in resonance peaks in the cross sections. Near these resonance peak frequencies ω_p one may expand the dielectric function as

$$\epsilon(\boldsymbol{q},\omega + \mathrm{i}0^+) \approx 1 - \frac{\omega_p^2(\boldsymbol{q})}{(\omega + \mathrm{i}0^+)^2} \tag{D.15}$$

and find in the plasmon pole approximation that

$$\begin{aligned} \frac{1}{\epsilon(\boldsymbol{q},\omega + \mathrm{i}0^+)} &\approx \frac{\omega_p^2(\boldsymbol{q})}{(\omega + \mathrm{i}0^+)^2 - \omega_\mathrm{p}^2} \\ &= \frac{\omega_\mathrm{p}(\boldsymbol{q})}{2} \left[\frac{1}{\omega + \mathrm{i}0^+ - \omega_\mathrm{p}(\boldsymbol{q})} - \frac{1}{\omega + \mathrm{i}0^+ + \omega_\mathrm{p}(\boldsymbol{q})} \right], \\ \Rightarrow \Im \frac{1}{\epsilon(\boldsymbol{q},\omega + \mathrm{i}0^+)} &= -\frac{\pi}{2} \omega_\mathrm{p}(\boldsymbol{q}) \left[\delta(\omega - \omega_\mathrm{p}(\boldsymbol{q})) - \delta(\omega + \omega_\mathrm{p}(\boldsymbol{q})) \right] \end{aligned} \tag{D.16}$$

Hence, at low temperatures (compared to $\hbar\omega_\mathrm{p}$ the cross section near the poles behaves as

$$\frac{\mathrm{d}\sigma_\mathrm{eels}}{\mathrm{d}\Omega_{\boldsymbol{q}}\mathrm{d}\omega} \propto \frac{k_1}{k_0} \tilde{u}(\boldsymbol{q}) \omega_\mathrm{p}(\boldsymbol{q}) \delta(\omega - \omega_\mathrm{p}(\boldsymbol{q})) \tag{D.17}$$

These relations can now be utilized for the interpretation of the electron-energy loss spectra (see Ref. [125] and references therein). Since the cross section is weighted with the form factor of the Coulomb potential, forward scattering is favored and hence plasmon modes with $q \approx 0$ are preferentially excited[1].

[1] We recall our conclusion of the preceding section that for small q the plasma mode is undamped whereas for large q the plasma frequencies coincide energetically with the p–h excitations and hence they decay. This is also reflected in the width of these peaks at the respective q. It is also worth noting that from a physical point of view, for $q = 0$ the plasma frequency corresponds to the eigenfrequency of the oscillating electron gas against the positive (neutralizing)ionic background. The driving force is the electric field that emerges at the boundaries due to charge-density build up. For $q \neq 0$ the driving force is created by the charge-density fluctuations of the electrons gas itself.

It should be mentioned in this context that in addition to electrons as a probing beam, a number of other sources, such as X-rays and neutrons have also been employed and similar formulas to Eq. (D.14) can be established for the corresponding cross sections [125].

D.2 Properties of the Pair-distribution Function

To illustrate explicitly some features of $g(\boldsymbol{r})$ and $S(\boldsymbol{q},\omega)$, as given by Eq. (D.12) let us consider the three-dimensional jellium model. The ground state is then the filled Fermi sphere and we readily verify that

$$\langle \epsilon_n | \hat{n}^\dagger_{\boldsymbol{q}} | \epsilon_0 \rangle = \begin{cases} N\delta_{n,0} & \text{if} \quad q = 0, \\ 1 & \text{if} \quad q \neq 0 \text{ and if } |\epsilon_n\rangle \text{ corresponds} \\ & \text{to p–h excitations present in } \hat{n}^\dagger_{\boldsymbol{q}}, \\ 0 & \text{else.} \end{cases} \tag{D.18}$$

Hence, we deduce for the dynamic structure factor the relation

$$\begin{aligned} S(\boldsymbol{q} \neq 0, \omega) &= \frac{4\pi}{N} \sum_{\boldsymbol{k}} \Theta(k_{\mathrm{F}} - k)\,[1 - \Theta(k_{\mathrm{F}} - |\boldsymbol{k}+\boldsymbol{q}|)]\,\delta(\omega - (\epsilon_{\boldsymbol{k}+\boldsymbol{q}} - \epsilon_{\boldsymbol{k}})/\hbar), \\ S(\boldsymbol{q} = 0, \omega) &= 2\pi N \delta(\omega) \end{aligned} \tag{D.19}$$

The static structure factor then has the form

$$\begin{aligned} S(\boldsymbol{q}) &= 2\pi \left[1 - \frac{2}{N} \frac{V}{(2\pi)^3} \int \mathrm{d}^3\boldsymbol{k}\, \Theta(k_{\mathrm{F}} - k)\Theta(k_{\mathrm{F}} - |\boldsymbol{k}+\boldsymbol{q}|) \right], \\ &= 2\pi \left[1 - \frac{V}{3N\pi^2}\, \Theta(2k_{\mathrm{F}} - q) \left(k_{\mathrm{F}}^3 - \frac{3}{4} q k_{\mathrm{F}}^2 + \frac{1}{16} q^3 \right) \right] \end{aligned} \tag{D.20}$$

Recalling that $k_{\mathrm{F}}^3 = 3\pi^2 \frac{N}{V}$ we can write

$$S(\boldsymbol{q}) = 2\pi \left\{ 1 - \Theta(2k_{\mathrm{F}} - q) \left[1 - \frac{3q}{4k_{\mathrm{F}}} + \frac{q^3}{16k_F^3} \right] \right\} \tag{D.21}$$

For the particle density correlation function and the pair distribution function we also derive in a straightforward manner that

$$\begin{aligned} G(\boldsymbol{r}, 0) &= \bar{n} + \delta(\boldsymbol{r}) - \frac{2}{VN} \sum_{\boldsymbol{k}} \sum_{\boldsymbol{p}} \exp\left[-\mathrm{i}(\boldsymbol{p} - \boldsymbol{k}) \cdot \boldsymbol{r} \right] \Theta(k_{\mathrm{F}} - k)\Theta(k_{\mathrm{F}} - p), \\ &= \delta(\boldsymbol{r}) + \bar{n} g(\boldsymbol{r}) \end{aligned} \tag{D.22}$$

From this equation we can now infer the form of the pair distribution function for the three-dimensional electron gas, namely

$$\begin{aligned}
g(\boldsymbol{r}) &= 1 - \frac{2}{\bar{n}^2}\frac{1}{(2\pi)^6}\int \mathrm{d}^3\boldsymbol{p}\int \mathrm{d}^3\boldsymbol{k}\mathrm{e}^{-\mathrm{i}(\boldsymbol{p}-\boldsymbol{k})\cdot\boldsymbol{r}}\Theta(k_\mathrm{F}-k)\Theta(k_\mathrm{F}-p), \\
&= 1 - \frac{2}{\bar{n}^2}\left[\frac{i}{4\pi^2 r}\int_0^{k_\mathrm{F}} \mathrm{d}p\, p(\mathrm{e}^{-\mathrm{i}pr}-\mathrm{e}^{\mathrm{i}pr})\right]^2, \\
&= 1 - \frac{1}{2\pi^4\bar{n}^2 r^2}\left(\frac{\sin k_\mathrm{F} r}{r^2} - \frac{k_\mathrm{F} r\cos k_\mathrm{F} r}{r^2}\right), \\
&= 1 - \frac{k_\mathrm{F}^6}{2\pi^4\bar{n}^2}\left[\frac{\sin k_\mathrm{F} r - k_\mathrm{F} r\cos k_\mathrm{F} r}{k_\mathrm{F}^3 r^3}\right]^2, \\
&= 1 - \frac{9}{2}\left[\frac{\sin k_\mathrm{F} r - k_\mathrm{F} r\cos k_\mathrm{F} r}{k_\mathrm{F}^3 r^3}\right]^2
\end{aligned} \tag{D.23}$$

This relation reveals that the pair distribution function possesses a damped oscillatory behavior and acquires a minimal value (but does not vanish) in the limit of diminishing relative interelectronic distance, i.e.

$$\lim_{r\to 0} g(r) \to \frac{1}{2} \tag{D.24}$$

$$\lim_{r\to \infty} g(r) \to 1 \tag{D.25}$$

References

[1] J. P. Taylor, *Scattering Theory* (John Wiley & Sons, Inc., 1972).

[2] C. J. Joachain, *The Quantum Collision Theory* (North-Holland Publishing Company, 1983).

[3] J. Berakdar, *Concepts of Highly Excited Electronic Systems* (Wiley-VCH, Weinheim, 2003).

[4] E. Weigold, I. E. McCarthy, *Electron Momentum Spectroscopy* (Kluwer Academic/Plenum Publishers, 1999).

[5] M. Vos, A. S. Kheifets, E. Weigold, S. A. Canney, B. Holms, F. Aryasetiawan, K. Karlsson, *J. Phys.: Conden. Matter* **11**, 3545 (1999); E. Weigold, M. Vos, in *Many Particle Spectroscopy of Atoms, Molecules, Clusters, and Surfaces*, J. Berakdar and J. Kirschner (Eds.) (Kluwer Academic/Plenum Publishers, New York, 2001).

[6] S. Iacobucci, L. Marassi, R. Camilloni, S. Nannarone, G. Stefani, *Phys. Rev. B* **51**, 10252 (1995).

[7] C. N. Yang, *Phys. Rev.* **74**, 764 (1948).

[8] J. W. Cooper, R. N. Zare, *Lectures in Theoretical Physics llc* (Gordon and Breach, New York 1969) p. 317.

[9] J. A. R. Samson, *Philos. Trans. R. Soc. London, Ser. A* **268**, 141–6 (1970).

[10] T. E. H. Walker, J. T. Weber, *J. Phys. B: Atom. Molec. Phys.* **7**, 674–92 (1974).

[11] V Schmidt, *Phys. Lett. A* **45**, 634 (1973).

[12] P. A. M. Dirac, *Proc. R. Soc. London* **160**, 48 (1937).

[13] G. Garibotti, J. E. Miraglia, *Phys. Rev. A* **21**, 572 (1980).

[14] M. Brauner, J.S. Briggs, H. Klar, *J. Phys. B*, **22**, 2265 (1989).

[15] A. Sommerfeld, *Atombau und Spektrallinien* (F. Vieweg und Sohn, Braunschweig, 1939), II Band.

Electronic Correlation Mapping: From Finite to Extended Systems. Jamal Berakdar

ISBN: 3-527-40350-7

[16] C. Caroli, D. Lederer-Rozenblatt, B. Roulet, D. Saint-James, *Phys. Rev. B* **8**, 4552 (1973).

[17] M. Abramowitz, I. Stegun, *Pocketbook of Mathematical Functions*, (Verlag Harri Deutsch, Frankfurt,1984).

[18] S. D. Kevan (Ed.), *Angle-Resolved Photoemission: Theory and Current Application*, Studies in Surface Science and Catalysis (Elsevier, North Holland, 1992).

[19] S. Hüfner, *Photoelectron Spectroscopy*, No. 82 in Springer Series in Solid-State Science, (Springer Verlag, Heidelberg, 1995).

[20] *Solid-State Photoemission and Related Methods: Principles and Practices* M. A. van Hove (Editor), W. M. Schattke (Eds.) (Wiley, Berlin, 2003).

[21] G. Wannier, *Phys. Rev.* **90**, 817 (1953).

[22] R.K. Peterkop, *J. Phys. B* **4**, 513 (1971).

[23] A.R.P. Rau, *Phys. Rev. A* **4**, 207 (1971).

[24] H. Klar, *J. Phys. B* **14**, 3255 (1981).

[25] A.R.P. Rau, *Phys. Rep.* **110**, 369 (1984).

[26] J.M. Feagin, *J. Phys. B* **17**, 2433 (1984).

[27] A. Huetz, P. Selles, D. Waymel, J. Mazeau, *J. Phys. B* **24**, 1917 (1991).

[28] A.K. Kazansky, V.N. Ostrovsky, *Phys. Rev. A* **48**, R871 (1993).

[29] J-M. Rost, *J. Phys. B* **28**, 3003 (1995); *Phys. Rep.* **297**, 271 (1998).

[30] J.H. Macek, S. Yu. Ovchinnikov, *Phys. Rev. Lett.*, **74**, 4631 (1995).

[31] M.B. Shah, D.S. Elliott, H.B. Gilbody, *J. Phys. B*, **20**, 3501 (1987).

[32] S. Matt, B. Dünser, M. Lezius, K. Becker, A. Stamatovic, P. Scheier, T. D. Märk, *J. Chem. Phys.* **105** 1880 (1996).

[33] V. Foltin, M. Foltin, S. Matt, P. Scheier, K. Becker, H. Deutsch, T. D. Märk, *Chem. Phys. Lett.* **289**, 181 (1998).

[34] V. Tarnovsky, P. Kurunczi, S. Matt, T. D. Märk, H. Deutsch, K. Becker, *J. Phys. B* **31**, 3403 (1998).

[35] H. Deutsch, K. Becker, J. Pittner, V. Boniacic-Koutecy, S. Matt, T. D. Mark, *J. Phys. B* **29**, 5175 (1996); M. Lezius, P. Scheier, M. Foltin, B. Dünser, T. Rauth, V.M. Akimov, W. Krätschmer, T.D. Märk, *Int. J. Mass Spectrom. Ion Processes*, **12** 949–56(1993); S.

Matt, O. Echt, T. Rauth, B. Dünser, M. Lezius, A. Stamatovic, P. Scheier, T.D. Märk, Z. Phys., D, **40**, 389–394 (1997); B. Dünser, M. Lezius, P. Scheier, H. Deutsch, T.D. Märk, *Phys. Rev. Lett.* **74**, 3364–3367 (1995).

[36] O. Kidun, J. Berakdar, *Phys. Rev. Lett.* **87**, 263401 (2001).

[37] H. Deutsch, K. Becker, R. Basner, M. Schmidt, T.D. Märk, *J. Phys. Chem.* **102**, 8819–8826 (1998); H. Deutsch, K. Becker, S. Matt, T.D. Märk, *Int. J. Mass Spectrometry* **197**, 37–69 (2000); M. Probst, H. Deutsch, K. Becker, T.D. Märk, *Int. J. Mass Spectrom.* **206**, 13–25 (2001); B. Coupier, B. Farizon, M. Farizon, M.J. Gaillard, F. Gobet, N.V. de Castro Faria, S. Ouaskit, M. Carre, B. Gstir, G. Hanel, S. Denifl, L.Feketeova, P.Scheier, T.D.Märk, *Eur. Phys. J.*, D **20**, 459–468 (2002); B. Gstir, G. Hanel, J. Fedor, M. Probst, P. Scheier, N.J. Mason, T.D. Märk, *J. Phys. B* **35**, 2567–2574 (2002); U. Onthong, H. Deutsch, K. Becker, S. Matt, M. Probst, T.D. Märk, *Int. J. Mass Spectrom.* **214**, 53–6 (2002); S.Pal, J.Kumar, T.D.Märk, Differential, partial and total *J. Chem. Phys.* **120**, 4658–4663 (2004); S.Ptasinska, S.Denifl, P.Scheier, T.D.Märk, *J. Chem. Phys.* **120** 8505–8511 (2004); V. Grill, G. Walder, P. Scheier, M. Kurdel, T.D. Märk, *Int. J. Mass Spectrom. Ion Processes* **129**31–42 (1993).

[38] L. D. Landau, *Sov. Phys. JETP* **3**, 920 (1957).

[39] *Magnetism Beyond 2000*, A.J. Freeman, S.D. Bader (Eds.) (Elsevier, Amsterdam 1999).

[40] J. Kirschner, *Polarized Electrons at Surfaces* (Springer Verlag, Berlin,Heidelberg, New York, Tokyo,1985).

[41] H. Ibach, D. L. Mills *Electron Energy Loss Spectroscopy and Surface Vibrations* (Academic Press, New York, 1982).

[42] *Polarized Electrons in Surface Physics*, R. Feder (Ed.) (World Scientific, Singapore, 1985).

[43] J. Kirschner, D. Rebenstorff, H. Ibach, *Phys. Rev. Lett.* **53**, 698 (1984).

[44] M. Plihal, D.L. Mills, J. Kirschner, *Phys. Rev. Lett.* **82**, 2579 (1999).

[45] D. L. Abraham, H. Hopster, *Phys. Rev. Lett.* **62**, 1157 (1989).

[46] R. Vollmer, M. Etzkorn, P. S. Anil Kumar,1 H. Ibach, J. Kirschner, *Phys. Rev. Lett.* **91**, 147201 (2003).

[47] K. A. Kouzakov, J. Berakdar, *Phys. Rev. A* **68**, 022902/1–022902/8 (2003).

[48] J. Kessler, *Polarized Electrons* (Springer Verlag, Berlin, Heidelberg, New York, 1976).

[49] H. Th. Prinz, K. H. Besch, W. Nakel, *Phys. Rev. Lett.* **74**, 243 (1995).

[50] H. Ast, S. Keller, C. T. Whelan, H. R. J. Walters, R. M. Dreizler, *Phys. Rev. A* **50**, R1 (1994).

[51] S. Keller, C. T. Whelan, H. Ast, H. R. J. Walters, R. M. Dreizler, *Phys. Rev. A* **50**, 386 (1994).

[52] G. Baum, W. Blask, P. Freienstein, L. Frost, S. Hesse, W. Raith, P. Rappolt and M. Streun, *Phys. Rev. Lett.* **69**, 3037 (1992).

[53] M. Streun, G. Baum, W. Blask, J. Rasch, I. Bray, D. V. Fursa, S. Jones, D. H. Madison, H. R. J. Walters, C. T. Whelan, *J. Phys. B* **31**, 4401 (1998).

[54] B. Granitza, X. Guo, J. M. Hurn, J. Lower, S. Mazevet, I. E. McCarthy, Y. Shen, E. Weigold, Aust. J. Phys. **49**, 383 (1996)

[55] A. Dorn, A. Elliot, X. Guo, J. Hurn, J. Lower, S. Mazevet, I. E. McCarthy, Y. Shen, E. Weigold, *J. Phys. B* **30**, 4097 (1997)

[56] S. Mazevet, *PhD Thesis,* Australian National University, 1997.

[57] G. F. Hanne, *Can. J. Phys.* **74**, 811 (1996).

[58] C. Mette, T. Simon, C. Herting, G. F. Hanne, D. H. Madison, *J. Phys. B* **31**, 4689 (1998).

[59] D. H. Madison, V. D. Kravtsov, S. Mazevet, *J. Phys. B* **31**, L17 (1998).

[60] U. Fano, A.R.P. Rau, *Atomic Collisions and Spectra* (Academic Press, New York) p. 342.

[61] K. Blum, *Density Matrix Theory and Applications* (Plenum, New York, 1981).

[62] D.M. Brink, G.R. Satchler, *Angular momentum*, 2nd Edn. (Clarendon Press, Oxford 1968).

[63] D. A. Varshalovich, A. N. Moskalev, V. K. Khersonskii, *Quantum Theory of Angular Momentum* (World Scientific, Singapore,1988).

[64] J. Berakdar, P.F. O'Mahony, F. Mota Furtado, *Z. Phys. D*, **39**, 41 (1997).

[65] H. Gollisch, X. Yi, T. Scheunemann, R. Feder, *J. Phys.: Conden. Matter* **11**, 9555–9570 (1999); H. Gollisch, Feder R., *J. Phys.: Conden. Matter* **16**, 2207–2214 (2004), *Solid State Commun.* **119**,625–629, (2001).

[66] S. Samarin, O.M. Artamonov, A.D. Sergeant, J. Kirschner, A. Morozov, J.F.Williams, *Phys. Rev. B.* **7007**, 3403 (2004); S. Samarin, O.M.Artamonov, A.D. Sergeant, J.F. Williams, *Surf. Sci.*, **579**, 166–174 (2005).

[67] J. Berakdar, S. N. Samarin, R. Herrmann, J. Kirschner, *Phys. Rev. Lett.* **81**, 3535 (1998).

[68] J. Berakdar, M. P. Das, *Phys. Rev. A* **56**, 1403 (1997).

[69] G.D. Fletcher, M.J. Alguard, T.J. Gay, V.W. Hughes, P.F. Wainwright, M.S. Lubell, W. Raith, *Phys. Rev. A* **31**, 2854 (1985).

[70] Crowe, D.M., X.Q. Guo, M.S. Lubell, J. Slevin, M. Eminyan, *J. Phys. B* **23**, L325 (1990).

[71] C.H. Greene, A.R.P. Rau, *Phys. Rev. Lett.* **48**, 533 (1982).

[72] H. Ehrhardt, M. Schulz, T. Tekaat, K. Willmann, *Phys. Rev. Lett.* **22**, 89 (1969).

[73] U. Amaldi, A. Egioli, R. Marconero, G. Pizella, *Rev. Sci. Instrum.* **40**, 1001 (1969).

[74] B. Lohmann, I. E. McCarthy, A. T. Stelbovics, E. Weigold, *Phys. Rev. A* **30**, 758 (1984).

[75] H. Ehrhardt, K. Jung, G. Knoth, P. Schlemmer, *Z. Phys. D* **1**, 3 (1986).

[76] J. Berakdar, J. Röder, J. S. Briggs, H. Ehrhardt, *J. Phys. B* **29**, 6203 (1996).

[77] F. W. Byron Jr., C. J. Joachain, *Phys. Rep.* **179**, 211 (1989).

[78] I. E. McCarthy, E. Weigold, *Rep. Prog. Phys.* **54**, 781 (1991).

[79] J. S. Briggs, *Comments At. Mol. Phys.* **23**, 155 (1989).

[80] T. N. Rescigno, M. Baertschy, W. A. Isaacs, C. McCurdy, *Science* **286**, 2474 (1999).

[81] M. Baertschy, T. N. Rescigno, C. W. McCurdy, *Phys. Rev. A* **64**, 022709 (2001).

[82] M. Baertschy, T. N. Rescigno, W. A. Isaacs, X. Li, C. W. McCurdy, *Phys. Rev. A* **63**, 022712 (2001).

[83] Ph. L. Bartlett, A. T. Stelbovics, I. Bray, *J. Phys. B* **37** L69-L76 (2004).

[84] I. Bray, A. T. Stelbovics, *Phys. Rev. A* **46**, 6995 (1992); I. Bray, D. V. Fursa, *Phys. Rev. A* **54**, 2991 (1996); I. Bray, *Phys. Rev. Lett.* **78**, 4721 (1997); A. Kheifets, I. Bray, *J. Phys. B* **31**, L447 (1998).

[85] I. Bray, *J. Phys. B*, **36** 2203 (2003).

[86] D. Proulx, R. Shakeshaft, *Phys. Rev. A* **48**, R875 (1993); M. Pont, R. Shakeshaft, *J. Phys. B* **28**, L571 (1995).

[87] J. Colgan, M. S. Pindzola, F. Robicheaux, *J. Phys. B* **34**, L457 (2001).

[88] L. Malegat, P. Selles, A. Kazansky, *Phys. Rev. Lett.* **85**, 4450 (2000).

[89] J. Briggs, V. Schmidt, *J. Phys. B*, **33**, R1 (2000).

[90] G. C. King, L. Avaldi, *J. Phys. B*, **33**, R215 (2000).

[91] L. Malegat, P. Selles, A. Huetz, *J. Phys. B* **30**, 251 (1997).

[92] *Many-particle Quantum Dynamics in Atomic and Molecular Fragmentation*, J. Ullrich, V. Shevelko (Eds.) (Springer, Berlin, 2003).

[93] M. A. Coplan, J.H. Moore, J.P.Doering, *Rev. Mod. Phys.* **66**(3), 985–1014, (1994); M. Vos, I.E. McCarthy, *Rev. Mod. Phys.* **67**(3), 713–723, (1995); W. Nakel, C.T. Whelan, *Phys. Rep.* **315**(6), 409–471 (1999); J. Berakdar, A. Lahmam-Bennani, C. Dal Cappello, *Phys. Rep.* **374**, 91–164 (2003); R. Dörner, V. Mergel, O. Jagutzki, L. Spielberger, J. Ullrich, R. Moshammer, H. Schmidt-Böcking, *Phys. Rep.* **330**(2–3), 96–192 (2000); J. Ullrich, R. Moshammer, A. Dorn, R. Dörner, L.P.H. Schmidt, H. Schmidt-Böcking, *Rep. Prog. Phys.* **66**, 1463 (2003).

[94] D. H. Madison, R. V. Calhoun, W. N. Shelton, *Phys. Rev. A* **16**, 552 (1977).

[95] F. Rouet, R. J. Tweed, J. Langlois, *J. Phys. B* **29**, 1767 (1996).

[96] E. P. Curran, H. R. J. Walters, *J. Phys. B* **20**, 337 (1987).

[97] E. P. Curran, C. T. Whelan, H. R. J. Walters, *J. Phys. B* **24**, L19 (1991).

[98] I. Bray, D. A. Konovalov, I. E. McCarthy, A. T. Stelbovics, *Phys. Rev. A* **50**, R2818 (1994).

[99] M. Brauner, J. S. Briggs, H. Klar, J. T. Broad, T. Rösel, K. Jung, H. Ehrhardt, *J. Phys. B* **24**, 657 (1991).

[100] J. Berakdar, H. Klar, M. Brauner, J. S. Briggs, *Z. Phys. D* **16**, 91 (1990).

[101] H. Klar, *Z. Phys. D* **16**, 231 (1990).

[102] J.S. Briggs, *Phys. Rev. A* **41** 539 (1990).

[103] J. Berakdar, *Phys. Rev. A* **53**, 2314 (1996).

[104] H. Klar, D. A. Konovalov, I. E. McCarthy, *J. Phys. B* **26**, L711 (1993).

[105] S. Jones, D. H. Madison, A. Franz, P. L. Altick, *Phys. Rev. A* **48**, R22 (1993).

[106] S. Jones, D. H. Madison, D. A. Konovalov, *Phys. Rev. A* **55**, 444 (1997).

[107] J. Berakdar, *Phys. Rev. A* **55**, 1994–2003 (1997).

[108] J. Berakdar, J.S. Briggs, *Phys. Rev. Lett.*, **72**, 3799 (1994); *J. Phys. B* **27**, 4271 (1994).

[109] E. O. Alt, A. M. Mukhamedzhanov, *Phys. Rev. A* **47**, 2004 (1993).

[110] J. Berakdar, *Aust. J. Phys.* **49**, 1095 (1996).

[111] J. Berakdar, J.S. Briggs, I. Bray, D.V. Fursa, *J. Phys. B* **32**, 895 (1999).

[112] M. Streun, G. Baum, W. Blask, J. Berakdar, *Phys. Rev. A* **59**, R4109-R4112 (1999).

[113] J. Berakdar, *Phys. Rev. A* **54** 1480 (1996).

[114] Sh.D. Kunikeev, V.S. Senashenko, Zh. E′ksp. Teo. Fiz. **109**, 1561 (1996); *Nucl. Instrum. Methods B*, **154**, 252 (1999).

[115] G. Gasaneo, F.D. Colavecchia, C.R. Garibotti, J.E. Miraglia, P. Macri, *Phys. Rev. A* **55**, 2809 (1997); G. Gasaneo, F.D. Colavecchia, C.R. Garibotti, *Nucl. Instrum. Methods B* **154**, 32 (1999).

[116] J. Berakdar, *Phys. Lett. A* **220**, 237 (2000); *Phys. Lett. A*, **277**, 35 (1997).

[117] E.O. Alt, A.M. Mukhamedzhanov, *Phys. Rev. A* **47**, 2004 (1993); E.O. Alt, M. Lieber, *Phys. Rev. A*, 54, 3078 (1996).

[118] J. Röder, J. Rasch, K. Jung, C. T. Whelan, H. Ehrhardt, R. J. Allan, J. H. R. Walters, *Phys. Rev. A* **53**, 225 (1996).

[119] J. Rasch, C. T. Whelan, R. J. Allan, S. P. Lucey, H. R. J. Walters, *Phys. Rev. A* **56** 1379–83 (1997).

[120] A. J. Murray, N. J. Bowring, F. H. Read, *J. Phys. B* **33**, 2859–2867 (2000); A. J. Murray, F. H. Read, *J. Phys. B* **33**, L297-L302 (2000); N. J. Bowring, F. H. Read, A. J. Murray, *J. Phys. B* **32** L57-L63 (1999).

[121] O. Samardzic, J.A. Hurn, E. Weigold, M.J. Brunger, *Aust. J. Phys.* **47**, 703–720 (1994); D.H.Madison, V.D. Kravtsov, J.B. Dent, M. Wilson, *Phys. Rev. A* **56**, 1983–1988 (1997); V.V.Balashov, *J. Phys. IV* **9**(P6), 21–24 (1999); R. Flammini, E. Fainelli, L. Avaldi, *J. Phys. B* **33**,1507–1519 (2000); D.A. Horner, C.W. McCurdy, T.N. Rescigno, *Phys. Rev. A* **7101**(1), 701 (2005).

[122] J. Lower, E. Weigold, J. Berakdar, S. Mazevet, *Phys. Rev. Lett.* **86**, 624–627 (2001); P. Golecki, H. Klar, *J. Phys. B* **32**, 1647–1656 (1999).

[123] C. Hohr, A. Dorn, B. Najjari, D. Fischer, C.D. Schröter, J. Ullrich, *Phys. Rev. Lett.* **9415**, 3201 (2005); C. J. Joachain, P. Francken, A. Maquet, P. Martin, V. Veniard, *Phys. Rev. Lett.* **11**, 165 (1988); P. Martin, V. Veniard, A. Maquet, P. Francken, C. J. Joachain,

Phys. Rev. A **19**, 6178 (1989); F. Ehlotzky, A. Jaroń, J. Z. Kamiński, *Phys. Rep.* **297**, 63 (1998); F. Ehlotzky, *Phys. Rep.* **345**, 175 (2001); A. Makhoute, D. Khalil, A. Maquet, R. Taïb, *J. Phys. B*: At. Mol. Opt. Phys. **32**, 3255 (1999). S.-M. Li, J. Berakdar, J. Chen, Z.-F. Zhou, *J. Phys. B* **38**, 1291–1303 (2005).

[124] P. Fulde, *Electron Correlation in Molecules and Solids*, Springer Series in Solid-State Sciences, Vol. 100, (Springer Verlag, Berlin, Heidelberg, New York, 1991).

[125] Mahan G. D., *Many-Particle Physics*, 2nd edn., (Plenum Press, London 1993).

[126] G. Kotliar, D. Vollhardt, *Physics Today* **57**, 53–59 (2004).

[127] A. Moreo, S. Yunoki, E. Dagotto, *Science* **283**, 2034–2040 (1999).

[128] D. S. Saraga, B. L. Altshuler, D. Loss, R. M. Westervelt *Phys. Rev. Lett.* **92**, 246803–246807 (2004).

[129] R.W. Hill, C. Proust, L. Taillefer, P. Fournier, R.L. Greene, *Nature* **414**, 711–715 (2001).

[130] M. Uehara, S. Mori, C.H. Chen, S.W. Cheong, *Nature* **399**, 560–563 (1999).

[131] Y. Ji, M. Heiblum, D. Sprinzak, D. Mahalu, H. Shtrikman, *Science* **290**, 779–783 (2000).

[132] *Many-Electron Densities and Reduced Density Matrices* Series: Mathematical and Computational Chemistry, J. Cioslowski (Ed.), (Springer, Berlin, 2000).

[133] R. G. Parr, W. Yang, *Density Functional Theory of Atoms and Molecules* (Oxford University Press, New York, 1989).

[134] J. C. Slater, *Introduction to Chemical Physics* (McGraw-Hill, New York, 1939).

[135] J. C. Slater, *Phys. Rev.* **81**, 385–390 (1951).

[136] E. Wigner, *Trans. Faraday Soc.* **34**, 678–685 (1938).

[137] M. Nekovee, W.M.C. Foulkes, R.J. Needs, *Phys. Rev. Lett.* **87**, 036401–036404 (2001).

[138] G. Ortiz, I. Souza, R.M. Martin, *Phys. Rev. Lett.* **80**, 353–356 (1998).

[139] J.P. Perdew, K. Burke, Y. Wang, *Phys. Rev. B-Condensed Matter* **54**, 16533–16539, (1996).

[140] P.H.Acioli, D.M. Ceperley, *Phys. Rev. B-Conden. Matter* **54**, 17199–17207 (1996).

[141] W.E.Pickett, J.Q.Broughton, *Phys. Rev. B-Conden. Matter* **48**, 14859–14867 (1993).

[142] J.P.Perdew, J.A. Chevary, S.H. Vosko, K.A. Jackson, M.R. Pederson, D.J. Singh, C. Fiolhais, *Phys. Rev. B-Conden. Matter* **46**, 6671–6687 (1992).

[143] J.E. Inglesfield, J.D. Moore, *Solid State Commun.* **26**, 867–8671 (1978).

[144] O. Gunnarsson, B.I. Lundqvist, *Phys. Rev. B* **13**, 4274–4298 (1976).

[145] R. G. Parr, W. Yang, *Density-Functional Theory of Atoms and Molecules* (Oxford University Press, Oxford, 1989).

[146] A. L. Fetter, J. D. Walecka, *Quantum Theory of Many-particle Systems*, (New York, McGraw-Hill, 1971).

[147] H. Eschrig, *The Fundamentals of Density Functional Theory* (Teubner, Stuttgart, 1996).

[148] I. Lindgren, J. Morrison, *Atomic Many-Body Theory* (Springer Verlag, Berlin 1982).

[149] R. M. Dreizler, E. K. U. Gross, *Density Functional Theory* (Springer, Berlin, 1995).

[150] J. F. Dobson, G. Vignale, M. P. Das (eds.), *Electron Density Functional Theory, Recent Progress and New Directions*, (Plenum Press, 1998).

[151] A. Gonis, *Theoretical Materials Science: Tracing the Electronic Origins of Materials Behavior*, (Materials Research Society, Warrendale, 2000).

[152] C. Møller, K. Dan, *Vidensk. Selsk. Mat. Fys. Medd.* 23, 1 (1945).

[153] B. A. Lippmann, J. Schwinger, *Phys. Rev.* **79**, 469 (1950).

[154] F. O. Schumann, J. Kirschner, J. Berakdar, *Phys. Rev. Lett.* **95** 117601 (2005).

[155] S. Samarin, O. Artamonov, J. Berakdar, A. Morozov, J.Kirschner, *Surf. Sci.* **482**(2), 1015–1020 (2001). A. Morozov et al., *Phys. Rev. B* **6510**,(10) 4425 (2002).

[156] S. Samarin, J. Berakdar, A. Morozov, J.Kirschner, *Phys. Rev. Lett.* **85**, 1746 (2000).

[157] A. Morozov, J. Berakdar, S. N. Samarin, F. U. Hillebrecht, J. Kirschner, *Phys. Rev. B* **65**, 104425 (2002).

[158] I. Shavitt, *The Method of Configuration Interaction*, in Methods of Electronic Structure Theory, H. F. Schaefer (Ed.), (Plenum Press, New York, 1977), pp. 189–275.

[159] Recent Advances in Computational Chemistry – Vol. 3 *Recent Advances In Coupled-Cluster Methods* R. J. Bartlet (Ed.) (World Scientific, Singapore 2002).

[160] J. Cizek, *Adv. Chem. Phys.* 1435 (1969).

[161] J. Gauss, D. Cremer *Analytical Energy Gradients in Møller-Plesset Perturbation and Quadratic Configuration Interaction Methods: Theory and Application*, Advances in Quantum Chemistry Vol. **23**, 205–299 (1992).

[162] A. F. Hebard et al., *Nature* **350** 600 (1991).

[163] R. W. Lof et al., *Phys. Rev. Lett.* **68** 3924 (1992); *J. Electron. Spectrosc. Relat. Phenom.* **72** 83 (1995).

[164] D. P. Joubert, *J. Phys.: Condens. Matter* **5** 8047 (1993).

[165] O. Gunnarsson, G. Zwicknagel, *Phys. Rev. Lett.* **69**, 957 (1992).

[166] A.I. Liechtenstein, O.Gunnarsson, M. Knupfe, J. Fink, J.F. Armbruster *J. Phys.: Condens. Matter* **8**, 4001 (1996).

[167] K.-D. Tsuei, J.Y. Yuh, C.-T. Tzeng, R.-Y. Chu, S.-C. Chung, K.-L. Tsang, *Phys. Rev. B* **56**, 15412 (1997).

[168] F.U. Hillebrecht, A. Morozov, J. Kirschner, *Phys. Rev. B* **71**, 125406/1–6 (2005).

[169] R. Saito, G. Dresselhaus, M.S. Dresselhaus, *Physical Properties of Carbon Nanotubes* (World Scientific, Singapore, 1998).

[170] P. J. F. Harris, *Carbon Nanotubes and Related Structures: New Materials for the Twenty-First Century* (Cambridge University Press, Cambridge 1999).

[171] S. Reich, C. Thomsen, J. Maultzsch, *Carbon Nanotubes* (Wiley-VCH, Berlin, 2004).

[172] S. Keller, E. Engel, *Chem. Phys. Lett.* **299**, 165 (1999).

[173] M. Vos, S.A. Canney, I.E. McCarthy, S. Utteridge, M.T. Michalewicz, E. Weigold, *Phys. Rev. B* **56**, 1309 (1997).

[174] S. Keller, *Eur. Phys. J. D* **13**, 51 (2001).

[175] O. Kidun, J. Berakdar, *Surf. Sci.* **507–510** , 662 (2002).

[176] O. Kidun, N. Fominykh, J. Berakdar, *J. Phys. A* **35**, 9413 (2002).

[177] W. A. de Heer, *Rev. Mod. Phys.* **65**, 611–75 (1993).

[178] M. Brack, *Rev. Mod. Phys.* **5**, 677 (1993).

[179] F.U. Hillebrecht, A. Morozov, J. Kirschner, *Phys. Rev. B* **71**, 125406/1–6 (2005).

[180] A. Reinköster, U. Werner, N. M. Kabachnik, H. O. Lutz, *Phys. Rev. A* **64**, 23201 (2001).

[181] D. Hathiramani, P. Scheier, H. Bräuning, R. Trassl, E. Salzborn, L.P. Presnyakov, A.A. Narits, D.B. Uskov, *Nucl. Instrum. Methods in Phys. Res. B* **212**, 67 (2003).

[182] O. Kidun, N. Fominykh, J. Berakdar, *Chem. Phy. Lett.* **410**, 293–297 (2005).

[183] R. Camilloni, A. Giardini Guidoni, R. Tiribelli, G. Stefani, *Phys. Rev. Lett.* **29**, 618 (1972).

[184] J. Kirschner, O. M. Artamonov, A. N. Terekhov, *Phys. Rev. Lett.* **69**, 1711 (1992); O. M. Artamonov, S. N. Samarin, J. Kirschner, *Phys. Rev. B* **51**, 2491 (1992).

[185] J. Kirschner, O. M. Artamonov, S. N. Samarin, *Phys. Rev. Lett.* **75**, 2424 (1995).

[186] O. M. Artamonov, S. N. Samarin, J. Kirschner, *Appl. Phys. A* **65**, 535 (1997).

[187] O. M. Artamonov, S. N. Samarin, J. Kirschner, *Appl. Phys. A* **A65**, 535 (1997).

[188] R. Feder, H. Gollisch, D. Meinert, T. Scheunemann, O. M. Artamonov, S. N. Samarin, J. Kirschner, *Phys. Rev. B* **58**, 16418 (1998).

[189] A. S. Kheifets, S. Iacobucci, A. Ruocco, R. Camiloni, G. Stefani, *Phys. Rev. B* **57**, 7360 (1998).

[190] R. Feder, H. Gollisch, D. Meinert, T. Scheunemann, O. M. Artamonov, S. N. Samarin, and J. Kirschner, *Phys. Rev. B* **58**, 16418 (1998); H. Gollisch, T. Scheunemann, R. Feder, *Solid State Commun.* **117**(12) 691 (2001).

[191] J. Kirschner, O. M. Artamonov, and S. N. Samarin, *Phys. Rev. Lett.* **75**, 2424 (1995); O. M. Artamonov, S. N. Samarin, *Tech. Phys.* **46**, 1179 (2001).

[192] K.A. Kouzakov, J. Berakdar, *Phys. Rev. B* **66**, 235114 (2002); *J. Phys.: Condens. Matter* **15**, L41-L47 (2003).

[193] H.L. Skriver, *The LMTO Method* (Springer, Berlin, 1984).

[194] A. J. Freeman, *J. Magn. Magn. Mater.* **35**, 31 (1983).

[195] P. Blaha, K. Schwarz, P. Sorantin, S. B. Trickey. *Comput. Phys. Commun.* **59**, 399 (1990).

[196] L. Hedin, B. I. Lundqvist, *Solid State Phys.* **23**, 1 119 (1969).

[197] J. B. Pendry, *Low Energy Electron Diffraction* (Academic Press, London, 1974).

[198] J. F. L. Hopkinson, J. B. Pendry, D. J. Titterington, *Comput. Phys. Commun.* **19**, 69 (1980).

[199] R. H. Williams, G. P. Srivastava, I. T. McGovern, *Rep. Prog. Phys.* **43**, 1357 (1980).

[200] J. W. Krewer, Dissertation, Universität Duisburg, 1990.

[201] J. B. Pendry, *Photoemission and the Electronic Properties of Surfaces* B. Feuerbacher, B. Fitton, R. Willis (Eds.), (New York, Wiley, 1978) p. 87.

[202] M. A. van Hove, W. H. Weinberg, C.-M. Chan, *Low Energy Electron Diffraction, Springer Series in Surface Science* (Springer, Berlin, 1986).

[203] J. Berakdar, *Phys. Rev. Lett.* **83**, 5150 (1999).

[204] S. N.Samarin, J. Berakdar, O. Artamonov, H. Schwabe, J. Kirschner, *Surf. Sci.* **470**(1–2), 141–148 (2000); S. N.Samarin, J. Berakdar, A. Suvorova, O.M. Artamonov, D.K. Waterhouse, J. Kirschner, J.F. Williams, *Surf. Sci.* **548**(1–3), 187–199 (2004).

[205] P.C. Gibbons, S.E. Schnatterly, J.J. Ritsko, R. Fields, *Phys. Rev. B* **6**, 2451 (1976).

[206] V.U. Nazarov, *Phys. Rev. B* **56**, 2198 (1997).

[207] J. Lindhard, K. Dan, *Vidensk. Selsk. Mat. Fys. Medd.* **28**, No. 8 (1954); A.J. Glick, R.A. Ferrell, *Ann. Phys. (N.Y.)* **11**, 359 (1960).

[208] W. Kohn, N. Rostoker. *Phys. Rev.* **94**, 1111 (1954).

[209] J. Korringa, *Physica* **13**, 392 (1947).

[210] P. Weinberger, *Electron Scattering Theory for Ordered and Disordered Matter* (Clarendon Press, Oxford, 1990).

[211] N. Fominykh, J. Henk, J. Berakdar, P. Bruno, H. Gollisch, R. Feder, *Solid State Commun.*, **113**, 665 (2000).

[212] N. Fominykh, Berakdar, J. Henk, P. Bruno, *Phys. Rev.* Lett. **89** 086402, (2002).

[213] J. Berakdar, H. Gollisch, R. Feder, *Solid State Commun.*, **112**, 587 (1999).

[214] R. Feder, *J. Phys. C: Solid State Phys.* **14**, 2049 (1981).

[215] V. L. Moruzzi, J. F. Janak, A. R. Williams, *Calculated Electronic Properties of Metals* (Pergamon Press, New York, 1978).

[216] B. Segall, *Phys. Rev.* **105**, 108 (1957).

[217] K. Kambe, *Z. Phys. A* **22**, 32 (1967).

[218] K. Kambe, *Z. Phys. A* **22**, 422(1967).

[219] I. Turek, V. Drchal, J. Kudrnovský, M. Šob, P. Weinberger, *Electronic Structure of Disrodered Alloys, Surfaces, and Interfaces*, (Kluwer Academic, Boston, London, Dordrecht, 1997).

[220] A. Gonis, *Green Functions for Ordered and Disordered Systems* (North-Holland, Amsterdam, 1992); P. Weinberger, *Electron Scattering Theory for Ordered and Disordered Matter* (Clarendon, Oxford, 1990).

[221] B. L. Gyorffy, *Phys. Rev. B* **5**, 2382 (1972).

[222] J. Kudrnovský, V. Drchal, J. Măsek, *Phys. Rev. B* **35**, 2487 (1987).

[223] G. M. Stocks, H. Winter, *Z. Phys. B* **46**, 95 (1982).

[224] E. G. McRae, C. W. Caldwell, *Surf. Sci.* **57**, 77(1976).

[225] E. Tamura, R. Feder, *Solid State Commun.* **58**, 729 (1986).

[226] R. O. Jones, P. J. Jennings, O. Jepsen, *Phys. Rev. B* **29**, 6474 (1984).

[227] R. O. Jones, P. J. Jennings, *Surf. Sci. Rep.* **9**, 165 (1988).

[228] R. O. Jones, P. J. Jennings, M. Weinert, *Phys. Rev. B* **37**, 6113 (1988).

[229] E. Tamura, R. Feder, *Z. Phys. B* **81**, 425 (1990).

[230] J. Berakdar, A. Ernst, K.A. Kouzakov, *Nucl. Instrum. Methods Phys. Res. B* **233**, 125–131 (2005).

[231] R. Kubo, *J. Phys. Soc. Jpn.* **12**, 570 (1957).

[232] F. Stern, *Phys. Rev. Lett.* **18**, 546 (1967).

Index

h_{xc}, *see* exchange and correlation hole
2C approximation 12, 40, 62, 63
3C wave function 12, 38, 50, 56, 63, 68
3DWBA, *see* distorted-wave Born approximation

adiabatic connection 75
adiabatic switching 75
alignment 29
alkali metals 85
aluminum 112
angular distribution of photoelectrons 6
asymmetry parameter 6
asymptotic final state 15
ATA, *see* average t-matrix approximation
autoionizing states 71
average t-matrix approximation 136, 139

back-reflection geometry 103
band index 13
Bethe-ridge 136, 137
binary alloy 24, 140, 142, 144
binary peak 3, 5, 6
bipolar spherical harmonics 44
Bloch theorem 106, 155
Bloch vector 14
Boltzmann 21, 170
Born approximation 5, 7
Born series 43
Bragg condition 105
Brillouin zone 107, 155
BZ, *see* Brillouin zone

C_{60} 85, 86, 91–93, 95, 98, 99, 149
C–C bond length 91
canonical ensemble 24
Caroli formula 10
CC, *see* close-coupling scheme
CCC, *see* convergent close coupling
center-of-mass wave vector 102, 106, 107
charge susceptibility 81, 165, 170
charge-density response 95, 97, 99
charge-density-density correlation function 169
Clebsch–Gordon coefficient 27
close-coupling scheme 45
cluster 85, 90, 91, 95–97, 99
coherent potential approximation 136, 139
collision time 21, 22
colossal magnetoresistance 73
complex momenta 12
conditional probability density 75
configurational average 136, 138, 142, 143
configurationally averaged cross section 137
confluent hypergeometric function 12
contact interaction 21
convergent close coupling 45, 48, 52, 61, 62
coplanar asymmetric geometry 54, 61
coplanar geometry 51, 53
correlated two-particle emission 85, 123, 126
correlated two-particle energy distribution 132
correlation factor 75, 79
Coulomb correlation 75, 77, 78, 127
Coulomb gauge 162
Coulomb phase 46
Coulomb potential 1, 33, 39, 44, 46, 58–60, 69, 162, 170
Coulomb screened potential 16
Coulomb waves 12
coupling-constant 75
coupling-constant dependent hole 75
CPA, *see* coherent-potential approximation
cross sections 81

crystal momentum 13, 120
crystal potential 30, 33, 103, 106–108, 111, 113, 122, 127, 155, 156
crystallographic direction 24
Cu(001) 114–116, 119, 121, 122
cubic symmetry 157
cuprates 73

delocalized electron 18, 73
density matrix 25, 26, 82
density of state 14, 31, 32, 37, 39, 40, 80, 108, 110, 119, 129, 164
density operator 24, 26, 160, 167
density-functional theory 86, 95
DFT, *see* density-functional theory
DFT-LDA 86, 89, 95, 105, 124
dielectric function 123, 125, 163, 165, 169, 170, 189
dielectric response 23, 89
differential cross section 4, 11, 68, 92, 93, 149, 169
dipole approximation 6
dipole term 6
direct scattering 9, 32, 33, 120
disorder 136, 139
disorder-induced localization 136
dissipation 164
dissipation-fluctuation theorem 169
distorted-wave Born approximation 46
DOS, *see* density of state
DS3C, *see* dynamically screened 3C wave function
dynamical charge 48
dynamical screening 47, 63, 86, 98, 126
dynamical screening functions 129
dynamical structure factor 168, 169
dynamically screened 3C wave function 47–50, 52, 54–57, 61–63, 70

ECS, *see* exterior complex scaling
effective mass 22
electron flux density 1
electron gas 16, 22, 37, 108, 129, 162, 165, 166, 170, 172
electron ionizing collision 24
electron–crystal interaction 26
electron–electron coupling 73
electron–electron coupling constant 76
electron–electron interaction 7, 15, 16, 18, 20, 22, 24, 26, 40, 73, 75, 85–87, 91, 92, 95, 96, 126, 128, 132, 133, 191
electron–electron scattering 3
electron–hole pair 23, 87
electron-pair emission 23, 30, 43, 70
electron-removal probability 86
electronic correlation 12, 74, 75, 85, 130
electronic density distribution 2
elementary excitations 23
energy conservation 20, 21, 26, 57, 120
exchange and correlation hole 75
exchange and correlation potential 127
exchange effect 58, 83, 89, 90, 97–99, 110, 117, 119, 120, 123
exchange energy 76
exchange scattering 32–34
exchange-correlation energy 76
exchange-induced effects 55
exchange-split 30
exterior complex scaling 44

FBA, *see* first Born approximation
Fe(110) 121, 123
Fermi correlation 74, 75, 78
Fermi energy 19, 24, 108, 129, 140
Fermi hole 74, 78
Fermi liquid 22, 73
Fermi momentum 20
Fermi point 22
Fermi sphere, *see* Fermi surface
Fermi surface 20, 22, 164, 165
ferromagnetic layer 36
final-state density 16
final-state interactions 7
fine structure 24
finite difference method 45
finite-size effects 96
first Born approximation 5, 6, 8, 19, 38, 40, 43, 46, 56, 60, 62, 63, 65, 110
fixed-relative angle geometry 68
Fock expansion 9
form factor 3, 16, 18, 19, 32, 37, 39, 51, 57, 68, 69, 94, 115, 118, 170
forward scattering 18, 170
Friedel oscillation 126
frozen core 82
fullerene 18, 85, 86, 91, 92

fully-differential cross section 1

Gamma function 12
generalized susceptibility 161
GF, *see* Green's function
Green's function 26, 79, 80, 103, 132, 137, 142, 161
group velocity 155

Hartree potential 127
Hartree–Fock 74
Hartree–Fock equation 88
heavy targets 24
Heisenberg equation 159
helium 69–71
HF, *see* Hartree–Fock
high-energy excitation 23
highest occupied molecular orbital 92, 95, 98
hole state 87
HOMO, *see* highest occupied molecular orbital
hot electron 20, 21, 23
hydrogen 3, 4
hydrogenic orbitals 3
hydrogenic system 3

impact ionization 24, 46, 56, 97, 98
impurity 16
inelastic mean-free path 35, 140
initial momentum-distribution function 6
inverse photoemission 85
inversion symmetry 32
ionization potential 17, 43, 60, 91, 97
iron 31, 32

jellium model 107, 108, 110, 112, 162, 171

Kato cusp conditions 9
KKR, *see* Korringa–Kohn–Rostoker method
Kondo impurity 73
Kondo lattice systems 73
Korringa–Kohn–Rostoker method 127, 130, 139
Kubo formalism 87, 159

Laguerre basis 45
Landau theory 22, 73
laser pumping 71
lattice sites occupation 136
layer Korringa–Kohn–Rostoker method 128, 131, 132, 150, 151
layer-resolved pair emission cross section 35
LDA, *see* local-density approximation
LEED state, *see* low-energy electron diffraction
Legendre polynomial 6
Li cluster 97
Lindhard dielectric function 123
linear muffin-tin orbital 127, 130, 139
linear response 81, 86, 159, 163
Lippmann–Schwinger equation 45
LKKR, *see* layer Korringa–Kohn–Rostoker method
LMTO, *see* linear muffin-tin orbital
local environment effects 136
local-density approximation 86, 89, 95, 127, 142, 147
local-field effect 78
low-energy electron diffraction state 108, 127, 128, 130–132, 151
low-energy excitation 23
low-momentum Fourier components 3

magnetic film 36
magnetic moment 136
magnetic sensors 23
magnetic surface 30, 118
magnetic system 23
magnetism 73
magnetization direction 24
magneto resistance 23
magnon 23
majority band 23
many-electron system 2
mean kinetic energy 13
mean potential energy 13
minority band 23
modified dynamical Lindhard–Mermin dielectric function 123
Møller operator 79, 80
momentum conservation 19–22, 70, 82, 103, 113
momentum distribution function 5
momentum operator 3
momentum spectral density 9
momentum uncertainty 5
momentum-dependent response 90
momentum-distribution function 4, 5, 15

momentum-space pair density 79
momentum-space wave functions 12
Mott cross section 137
Mott–Hubbard insulator 85
muffin-tin zero 103

Ni(001) 114, 115
nonequilibrium Green's function 10
nonlocal density of states 10
nonlocal potential 89
nonlocal single-particle state density 13
nonlocal two-particle density 11
non-overlapping muffin-tin 103
nonspecular reflection 124, 126

off-diagonal elements 14
on-site molecular Coulomb integral 85
one-dimensional problems 12
one-electron atom 6
optical limit 6, 7
optical potential 127, 132, 141, 142
OPWA, *see* orthogonalized plane wave approximation
orbitally oriented target 71
orthogonalized plane wave approximation 48, 52–54
orthogonalized plane-wave approximation 48

pair diffraction 107, 120, 146, 148
pair-correlation function 76, 77, 82, 83
parity conservation 44
particle distribution function 22, 139
particle location-uncertainty 5
particle–hole excitation 88, 96, 165, 169
particle–hole pair 90, 93, 140, 164
Pauli matrices 25
Pauli principle 20–22, 30
peaking approximation 13
penetration depth 23
phase shifts 7
phase space 21, 116
photoemission 6, 14, 85, 137
photoionization 51
photoionization cross section 6
plane wave 2, 5, 7–9, 11, 12, 14, 16, 19
plane-polarized radiation 6
plane-wave approximation 37–40, 55, 57, 70
plane-wave final state 7
plasma frequency 110, 125, 166, 170
plasmon 90, 125, 126, 166, 170
plasmon pole approximation 170
plasmon wave vector 125
plasmon-assisted two-particle scattering 126
point symmetries 31
Poisson's equation 163
polarization vector 6
polarized electron 23–25, 30, 32, 77, 128, 150
polarized photon 6
polarizibility 161
polymeric materials 85
positional disorder 136
positron 91, 98, 99, 102, 120
post form 15, 80
prior form 15
probability density 10, 77, 156
probability flux density 81
propensity rule 63
proton 91, 99, 102, 124–126
PSCC, *see* pseudo-state close-coupling scheme
pseudo momentum 155
pseudo-state close-coupling scheme 43, 48
pseudostate 45
PWA, *see* plane-wave approximation

qp, *see* quasiparticles
quantum information 73
quasiparticle 22, 73
quasiparticle concept 73

radial transition dipole 7
random-phase approximation 86, 89, 98, 161
random-phase approximation with exchange 89, 91–93, 95–98, 149
re-normalization of the particle mass 22
reciprocal lattice 36, 78, 146
reciprocal lattice vector 14, 34
recoil peak 6, 7
recoil-ion momentum 1
reduced density matrix 82
reflection mode 137
retarded density–density correlation function 161
retarded Green's function 11
retarded polarizibility 162
retarded two-particle Green's function 160
Riccati–Bessel function 44

rotation transformation 45
RPA, *see* random-phase approximation
RPAE, *see* random-phase approximation with exchange
Rutherford cross section 169

scattering length 21, 106
scattering operator 80
scattering rate 19–21
screened electron–electron interaction 86
screening length 16, 18, 39, 93, 96, 109, 115, 129, 130, 164
self-consistent mean-field 87
self-interaction 75
short-range order 136
single branch linear dispersion 22
single-particle Green's function 11, 14
single-site approximation 136
singlet channel 63, 70, 78
singlet cross section 59, 60, 62, 67–70
singlet scattering 58–60, 64, 70
S matrix element 79, 80, 129
solid C_{60} 85
Sommerfeld parameter 47, 48
spectral function 22, 26, 137, 139, 140, 148
spectral representation 10
spectroscopic studies 14
specular reflection 124, 126
SPEELS, *see* spin-polarized electron-energy loss spectroscopy
spherical harmonics 3
spherical jellium 91, 96
spherical tensor 28, 29
spin asymmetry 29, 32, 35–39, 55, 56, 60, 62, 118
spin polarization vector 25, 36
spin waves 23
spin–flip 30, 32
spin–orbit interaction 24, 30
spin-averaged momentum-space two-particle density 82
spin-dependent cross section 26
spin-dependent scattering 23
spin-polarized electron-energy loss spectroscopy 23
spin-polarized homogeneous electron gas 37
spin-resolved cross section 31
spin-resolved detection 24
spin-resolved orbital 27

standard representation 25
standing flux scattering formulation 83
state multipole 27, 28, 34
static structure factor 168, 171
statistical correlation 136
statistical fluctuation 136
statistical operator 159
statistical tensor 27
Stoner excitation 23
Stoner spectrum 23
strong two-particle correlation 127
substitutional disordered binary alloy 136
sum rule 76
superconductivity 73
surface scattering potential 26
surface-deposited fullerene 85
susceptibility of the free Fermion gas 164

TDCS, *see* triply differential cross section
tensor operator 28
tensorial rank 29
TF, *see* Thomas–Fermi model
thin solid film 101
Thomas–Fermi 16, 108, 164, 165
Thomas–Fermi model 16
Thomas–Fermi potential 18
three-body final state 46, 49, 53
tight-binding approach 99
time-independent Hartree–Fock 87
time-reversed LEED state 130
T matrix element 15, 57, 71
T operator 80
total cross section 15–17, 38, 90, 95, 96, 98
transition dipole operator 51
transition operator 2
transition probability 81, 82
transition rate 81
transmission mode 136, 137
triplet channel 63, 70, 78
triplet cross section 38, 60–63, 65, 67, 70, 98, 110, 118
triplet scattering 31, 36, 37, 39, 55, 58–60, 62–65, 70, 118
triplet transition 31, 37
triply differential cross section 51, 52, 54, 61, 68, 69
tungsten 130, 132
two-electron on-site correlated terms 143
two-particle correlation 73, 78, 136

two-particle correlation functions 123
two-particle density matrix 74
two-particle final state 8
two-particle Green's function 10
two-particle probability density 74
two-particle spectroscopy 79
two-particle spectrum 23, 99
two-particle state 15
two-particle state density 11
two-particle wave function 15, 27, 127

unitary transformation 24

valence-band electron 13, 86, 91, 94
valence-band electron emission 13
variable-phase method 86, 89
VCA, *see* virtual crystal approximation
velocity-dependent product charge 47
virtual crystal approximation 136, 138–140
von-Laue diffraction 106
VPM, *see* variable-phase method

Wannier function 156
Wannier index 48
Wannier theory 41
Wick's theorem 161
work function 108, 139, 146, 169

X-ray absorption 85

Yukawa potential 39